Impact of Hydrocarbons on PE/PVC Pipes and Pipe Gaskets

Awwa Research Foundation

About the Awwa Research Foundation

The Awwa Research Foundation (AwwaRF) is a member-supported, international, nonprofit organization that sponsors research to enable water utilities, public health agencies, and other professionals to provide safe and affordable drinking water to consumers.

The Foundation's mission is to advance the science of water to improve the quality of life. To achieve this mission, the Foundation sponsors studies on all aspects of drinking water, including supply and resources, treatment, monitoring and analysis, distribution, management, and health effects. Funding for research is provided primarily by subscription payments from approximately 1,000 utilities, consulting firms, and manufacturers in North America and abroad. Additional funding comes from collaborative partnerships with other national and international organizations, allowing for resources to be leveraged, expertise to be shared, and broad-based knowledge to be developed and disseminated. Government funding serves as a third source of research dollars.

From its headquarters in Denver, Colorado, the Foundation's staff directs and supports the efforts of more than 800 volunteers who serve on the board of trustees and various committees. These volunteers represent many facets of the water industry, and contribute their expertise to select and monitor research studies that benefit the entire drinking water community.

The results of research are disseminated through a number of channels, including reports, the Web site, conferences, and periodicals.

For subscribers, the Foundation serves as a cooperative program in which water suppliers unite to pool their resources. By applying Foundation research findings, these water suppliers can save substantial costs and stay on the leading edge of drinking water science and technology. Since its inception, AwwaRF has supplied the water community with more than $300 million in applied research.

More information about the Foundation and how to become a subscriber is available on the Web at **www.awwarf.org.**

Impact of Hydrocarbons on PE/PVC Pipes and Pipe Gaskets

Prepared by:
Say Kee Ong, **James A. Gaunt**, **Feng Mao**, and **Chu-Lin Cheng**
Dept. of Civil, Construction and Environmental Engineering
Iowa State University , Ames, Iowa 50011

and

Lida Esteve-Agelet and **Charles R. Hurburgh**
Dept. of Agricultural and Biosystems Engineering
Iowa State University, Ames, Iowa 50011

Sponsored by:
Awwa Research Foundation
6666 West Quincy Avenue, Denver, CO 80235-3098

Published by:

Awwa Research Foundation

American Water Works Association

IWA Publishing

DISCLAIMER

This study was funded by the Awwa Research Foundation (AwwaRF). AwwaRF assumes no responsibility for the content of the research study reported in this publication or for the opinions or statements of fact expressed in the report. The mention of trade names for commercial products does not represent or imply the approval or endorsement of AwwaRF. This report is presented solely for informational purposes.

CONTENTS

TABLES

FIGURES

FOREWORD

The Awwa Research Foundation is a nonprofit corporation that is dedicated to the implementation of a research effort to help utilities respond to regulatory requirements and traditional high-priority concerns of the industry. The research agenda is developed through a process of consultation with subscribers and drinking water professionals. Under the umbrella of a Strategic Research Plan, the Research Advisory Council prioritizes the suggested projects based upon current and future needs, applicability, and past work; the recommendations are forwarded to the Board of Trustees for final selection. The foundation also sponsors research projects through the unsolicited proposal process; the Collaborative Research, Research Applications, and Tailored Collaboration programs; and various joint research efforts with organizations such as the U.S. Environmental Protection Agency, the U.S. Bureau of Reclamation, and the Association of California Water Agencies.

This publication is a result of one of these sponsored studies, and it is hoped that its findings will be applied in communities throughout the world. The following report serves not only as a means of communicating the results of the water industry's centralized research program but also a tool to enlist further support of the nonmember utilities and individuals.

Projects are managed closely from their inception to the final report by the foundation's staff and large cadre of volunteers who willingly contribute their time and expertise. The foundation serves a planning and management function and awards contracts to other institutions such as water utilities, universities, and engineering firms. The funding for this research effort comes primarily from the Subscription Program, through which water utilities subscribe to the research program and make an annual payment proportionate to the volume of water they deliver and consultants and manufacturers subscribe based on their annual billings. The program offers a cost-effective and fair method for funding research in the public interest.

A broad spectrum of water supply issues is addressed by the foundation's research agenda: resources, treatment and operations, distribution and storage, water quality and analysis, toxicology, economics, and management. The ultimate purpose of the coordinated effort is to assist water suppliers to provide the highest possible quality of water economically and reliably. The true benefits are realized when the results are implemented at the utility level. The foundation's trustees are pleased to offer this publication as a contribution toward that end.

David E. Rager
Chair, Board of Trustees
Awwa Research Foundation

Robert C. Renner, P.E.
Executive Director
Awwa Research Foundation

ACKNOWLEDGMENTS

The authors wish to thank the utility partners for their cooperation and contributions:

City of Ames, Clark Ave., Ames, IA, U.S.A., Stephen P. Salvo
EPCOR Water Services, 100065 Jasper Ave., Edmonton, AB, Canada, Trina Chugg, Andrea Low, and Heather Zarski

The authors of this report are indebted to the following organizations and individuals for their contributions of expertise and materials to this project:

Endot Industries, Inc., South Point, OH, U.S.A.
Cresline Plastic Pipe Co., Inc., Council Bluffs, IA, U.S.A.
Chevron Phillips Chemical Company LP, Performance Pipe Division, Plano, TX, U.S.A.
Griffin Pipe, 2601 Ninth Ave., Council Bluffs, IA, U.S.A., Brett Bleske and Kenneth J. Kemp
Hultec S&B Technical Products, 1300 East Berry St., Fort Worth, TX, U.S.A., J. Wayne Bird, Shah Rahman, and Bradford G. Corbett, Jr.
Iplex Pipelines Australia Pty Ltd, 35 Alfred Road, Chipping Norton, NSW 2115 Australia
Specified Fittings, 164 W. Smith Rd., Bellingham, WA, U.S.A.
Uni-Bell PVC Pipe Association, 2711 LBJ Freeway, Suite 1000, Dallas, TX, U.S.A., Craig Fisher and Steve Cooper

The authors wish to thank the AwwaRF project managers, Jian Zhang and Maureen Hodgins for their advice and comments The authors also appreciated the advice and comments of the Project Advisory Committee:

Glen R. Boyd, Ph.D., P.E., HDR Engineering, Inc., Bellevue, WA, U.S.A.
Stewart Burn, CSIRO, Highlett, VIC, Australia
Wayne Miller, P. Eng., Windsor Utilities Commission, Windsor, ON, Canada (retired)

The authors acknowledge the contributions of Insoo Joo to the work with Tyton® gaskets reported in Chapter 6.

EXECUTIVE SUMMARY

Plastic pipes have been used increasingly for the conveyance of drinking water in water distribution systems and, in many countries, plastic pipes are the preferred materials. Plastic materials commonly used for the distribution of drinking water include polyvinyl chloride (PVC), polyethylene (PE), and polybutylene (PB). One of the factors for the increased use of plastic pipes is the ease of installation and handling of plastic pipes. In addition, plastic pipes are highly durable and offer good resistance to the chemicals used in water treatment, such as chlorine. In many urban areas, plastic pipes may come into contact with contaminated soils as a result of leaks from underground storage tanks, chemical spills, and improper disposal of used chemicals. These pollutants from leaking storage tanks and contaminated soils can and have posed serious threats to the longevity and structural integrity of plastic pipes and elastomeric gaskets which, in turn, can affect the water quality in the distribution system. Although there are research studies and case studies documenting the permeation of organic compounds through plastic pipes and elastomeric gaskets, there is still a lack of understanding of the performance of PE and PVC pipe materials and elastomeric gaskets in hydrocarbon-contaminated soils commonly encountered under field conditions.

RESEARCH OBJECTIVES

The overall objective of this project was to study the impact of petroleum-based hydrocarbons on PE and PVC pipes, and elastomeric gaskets. The specific objectives of this project were:

i. Conduct a survey of water utilities to obtain information on the types of pipe materials and gasket materials used for mains and service connections, permeation incidents, and successful use of plastic pipes (PE/PVC) or gasketed pipes in contaminated areas.
ii. Conduct hydrocarbon permeation experiments using the pipe-drum or pipe-bottle apparatus for PE/PVC pipes under simulated environmental conditions.
iii. Develop predictive laboratory tests that may predict the performance of plastic pipe materials in soils contaminated with petroleum-based hydrocarbons.
iv. Conduct experiments under simulated environmental conditions to assess the performance and failure of gasket materials exposed to hydrocarbon contamination.

APPROACH

The research approach consisted of four major tasks: Task 1 – Survey and case histories, Task 2 - Laboratory material testing under simulated field conditions, Task 3 - Predictive pipe materials tests, and Task 4 - Gasket material tests.

In Task 1, a survey instrument was developed and sent to water utilities requesting information such as total miles of mains, miles of each type of pipe, total number of service connections and types of pipes, types of gasket materials, number of incidences of permeation of mains and service connections, and successful use of plastic pipes or any gasketed pipes in contaminated areas. In Task 2, information collected from the survey was used in the development of laboratory experiments that simulated field conditions. In this task, pipe-bottle apparatuses were developed to test permeation of PE and PVC pipes by benzene, toluene, trichloroethylene (TCE) and gasoline. In addition, pipe-bottle experiments were conducted using aqueous saturated solutions of benzene, toluene, TCE and gasoline. Three predictive tests: gravimetric sorption test, moving front test and near infrared spectroscopy were tested to assess their suitability as predictive test methods to assess susceptibility of pipes to permeation by hydrocarbons. Task 4 consisted of testing ductile iron pipes with five different gasket materials and PVC pipe with two different gasket materials in pipe-drum apparatuses. Contamination conditions included free product gasoline, 100% saturated aqueous gasoline solutions, and dilute aqueous solutions of gasoline. In this report, a 20% saturated aqueous solution of a solvent or gasoline, for example, refers to an aqueous solution with concentration equal to 20% of the concentration of a solution saturated with the solvent or gasoline. In this task, the effect of hydrostatic pressure on hydrocarbon permeation was also investigated.

CONCLUSIONS

A summary of the various tasks, major findings and recommendations is provided in Table E1. Details of the conclusions of the research are described below.

Survey and Case Histories

Eighteen percent of the total miles of mains reported were PVC, 16% were ductile iron, and 66% reported others, which included cast iron, steel asbestos-cement and concrete pipes. Only 0.18% of the total mains reported were PE pipes. With respect to the total number of service lines, PE and PVC pipe accounted for 6% and 5%, respectively.

The overall impact of permeation on the water industry is relatively small. Reports of permeation incidents involving potable water distribution are rare, with reports of one incident per 14,000 miles of mains and 0.9 incidents per million service connections in this survey.

Twenty-eight out of the 151 utilities (19%) reported having a protocol for selecting pipe and gasket materials for use in contaminated areas. Of the 28 utilities, 10 reported protocols specifying ductile (DI) pipe with resistant gaskets (either nitrile (NBR) or fluoroelastomer rubber (FKM); 8 reported protocols specifying DI with no mention of gasket type; 4 reported "no plastic", which could mean using DI in contaminated areas; 2 reported using steel, and 4 reported other approaches. No utility reported specifying plastic pipes in contaminated areas.

Data obtained from the survey was not sufficient to draw conclusions regarding the thresholds of contamination that would result in hydrocarbon permeation and require a utility to choose more resistant pipe or gasket materials.

Hydrocarbon Permeation of PE pipes

Laboratory experiments indicated that PE pipe is rapidly permeated by gasoline and solvents, whether in groundwater, in soil pore vapor, or as the free product. Benzene was the first of the BTEX compounds (benzene, toluene, ethylbenzene and xylene) to permeate PE pipe from water contaminated with gasoline, with the first detectable concentrations exceeding the EPA's 5 μg/L MCL for all experiments. Equations were derived to predict the concentration of contaminants in PE pipes for various external contaminant concentrations during continuous flow and after periods of stagnation. For a given contaminant in soil water, smaller pipes are more vulnerable

Table E.1 Summary of Tasks, Experiments, and Findings

Chapter	Description	Findings	Recommendations
Chapter 2	Survey and Case Histories	• Total miles of mains reported: 18% - PVC, 16% - ductile iron, 0.18% - PE, 66% - others (cast iron, steel, asbestos-cement, and concrete pipes). • No. of service connections reported: 6% - PE, 5% - PVC • One permeation incident per 14,000 miles of mains and 0.9 incidents per million service connections • 28 out of 151 utilities reported having a protocol for selecting pipe/gaskets for use in contaminated areas: 10 - DI pipe with resistant gaskets (either NBR or FKM); 8 - DI with no mention of gasket type; 4 - no plastic, 2 – steel, 4 – other approaches	
Chapter 3	Permeation of PE Pipes	• Gasoline permeates rapidly through PE pipes at all saturation concentrations and contaminated soils conditions • Permeation rate is a function of aqueous or vapor concentration of the hydrocarbons • Depending on hydrocarbon concentrations in soil water and soils and pipe size, water flow (no stagnation) may dilute benzene concentration to below MCL	• No level contamination at which PE pipe is resistant to permeation of gasoline, chlorinated solvents or aqueous solutions of gasoline • Corrective action should be taken
Chapter 4	Permeation of Ungasketed PVC pipes	• Premium gasoline (and, therefore, groundwater or soils saturated with gasoline) - no permeation of PVC pipes • Pure toluene, benzene and TCE – permeation of PVC pipes • saturated vapors of toluene, benzene and TCE – permeation of PVC pipes • Saturated aqueous solutions of benzene, toluene, or TCE – permeation of PVC pipe if activity > 0.60	• PVC pipes impervious to gasoline and gasoline-saturated solutions • PVC pipes permeated by saturated aqueous solutions of benzene, toluene, and TCE • Decontamination of PVC not feasible – must replace permeated pipe
Chapter 5	Predictive Tests (Sorption, Moving Front and Near Infrared (NIR) Reflectance)	• Of the three predictive tests (sorption, moving front and near infrared spectroscopy), the moving front test is the most suitable for PVC pipes and provides information on permeation rates • For single compound concentrations $\leq$40% (v/v) of benzene, toluene, or TCE in reference fuel, sorption (permeation) was insignificant and moving fronts were not formed. • No moving fronts were formed in $\leq$ 60% (v/v) saturated aqueous solutions of benzene, toluene, or TCE • Premium gasoline fortified with toluene - no moving front observed for $\leq$ 32.2% (v/v) total BTEX, moving fronts formed for $\geq$ 43.1% (v/v) total BTEX • No moving fronts for $\leq$40% (v/v) of multi-component BTEX mixtures in reference fuel. Fronts formed for $\geq$ 60% (v/v)	• PVC pipes can tolerate an activity of 0.40 (40% aqueous solutions of benzene, toluene and TCE) without permeation (without including safety factor and impact of other environmental conditions) • PVC pipes impervious to dilute aqueous solutions of benzene, toluene, and TCE, commonly encountered in the field • PVC pipes compatible with sum of individual BTEX activities $\leq$0.40 (activities are additive)

(continued)

Table E.1 Summary of Tasks, Experiments, and Findings (Cont'd)

Chapter	Description	Findings	Recommendations
Chapter 6	Permeation of Gasketed (Tyton®) DI Pipe	• Gaskets exposed to premium gasoline: EPDM permeated in 35 days, CR in 50 days, SBR in 50 days (4", benzene – avg. 2.72 mg/j/d), NBR in 124 days (4", benzene – 0.159 mg/j/d) (FKM (very low BTEX detected – probably leakage) • SBR gaskets permeated when exposed to gasoline-saturated water (210 days) and 50% saturated (240 days) • No permeation observed for SBR gasket in 20% and 5% saturated aqueous solutions of gasoline • No permeation of NBR gaskets observed for any level of gasoline contamination in water • No correlation between permeation and hydrostatic pressure in pipe	• SBR and NBR gaskets may be used (without exceeding benzene MCL) under all conditions of gasoline contamination with minimal average flow in pipe (no stagnation) • Under conditions of stagnation in contact with free product gasoline, NBR gaskets may be used (without exceeding benzene MCL). • SBR and NBR gaskets compatible with any level of gasoline in groundwater • NBR gaskets most cost effective choice when a gasket material resistant to gasoline is desired.
Chapter 7	Permeation of Gasketed (Rieber) PVC pipes	• Permeation of premium gasoline observed within 21 days for SBR (2", benzene avg. 0.73 mg/j/d) and NBR gaskets (2", benzene avg. 0.19 mg/j/d). • No permeation after 9 months for SBR and NBR gaskets in gasoline-saturated water • NBR rubber compound used for Rieber gaskets is about 3x more permeable than NBR rubber compound used in Tyton gaskets (75 W% vs 24% W% sorption).	• SBR or NBR gaskets may be used (without exceeding benzene MCL) in direct contact with gasoline with a minimal average flow in pipe • After 8 hours of stagnation in direct contact with gasoline, benzene MCL exceeded in pipes ≤ 24" using SBR and pipes ≤ 10" using NBR gaskets. • SBR or NBR gaskets compatible with any level of gasoline in groundwater (without exceeding MCL)

than larger pipes and the risk of exceeding EPA MCLs is much higher during periods of stagnation than during continuous flow. Laboratory experiments showed that, due to sorption, the organic content of soils can lower the concentration of hydrocarbon contaminants in the soil pore water. But once the organic adsorption capacity of the soil is exceeded, the organic content of the soil makes no difference in the permeation since the steady state permeation rates in PE pipes are dependent on the external bulk concentrations regardless of the soil types.

Permeation of PVC Pipe by Hydrocarbons

Pipe-Bottle Experiments

Pipe-bottle experiments showed that PVC pipes were rapidly permeated by pure benzene (20 days), toluene (16 days), and TCE (6.5 days). PVC pipes were also rapidly permeated by saturated vapors of benzene (31 days), toluene (28 days) and TCE (13 days). Pipe-bottle experiments also indicated that 100% aqueous saturated solutions of TCE and benzene permeated the PVC pipes at 168 and 250 days, respectively, while permeation by aqueous saturated solutions of toluene was predicted to occur after about 12 months of exposure. However, the pipe-bottle experiments indicated that PVC pipes were impervious to premium gasoline and gasoline-saturated water for over two years of exposure and, therefore, can be used in soils contaminated with gasoline, regardless of the level of contamination. An explanation for this resistance to permeation is that the activities of swelling compounds such as benzene, toluene, and xylene in premium gasoline were insufficient to soften the PVC material. The activity of total BTEX in gasoline for the pipe-bottle experiments was approximately 0.25.

Predictive Test – Gravimetric Sorption Test

Using the gravimetric sorption test, sorption of premium gasoline by PVC pipes was found to be negligible with less than 1% weight gain – implying that no permeation occurred. Sorption was insignificant for concentrations of toluene ≤ 40% (v/v) in NIST reference fuel – implying that no permeation of toluene occurred at these concentrations. With aqueous solutions of benzene, toluene, and TCE, the weight gains for 60% saturated aqueous solutions and below were less than 2% after 8 months of exposure. Sorption of the 20% saturated aqueous solutions was found to be statistically similar to that of the control experiment - implying that sorption was negligible for contaminant levels below 20% saturated aqueous solutions of benzene, toluene, and TCE.

Predictive Test – Moving Front Test

The moving front test developed in this study is a precise and rapid test that directly measures the progression of permeation caused by swelling in a PVC pipe. By measuring the moving front rates for 58 PVC pipe specimens from five different manufacturers in 48-hr toluene tests, the moving front test was found to be a useful tool for comparing the relative susceptibility of PVC pipes to permeation.

In experiments with PVC pipes exposed to various concentrations of benzene, toluene, and TCE in NIST reference fuel, no moving front was observed when the volume % of toluene was 40% or below. Moving fronts were observed at 100% and 80% saturated aqueous solutions of toluene, but no moving fronts at 60%, 40%, and 20% saturated aqueous solutions of benzene, toluene, or TCE after 12 months of exposure.

Moving fronts did not form during two years of exposure of PVC pipes to pure gasoline or to aqueous saturated solutions. Premium gasoline fortified with toluene did not cause a moving front to form in PVC pipe when the concentrations of total BTEX and toluene (v/v) were at or below 32.2% and 21.2%, respectively. Moving fronts were formed when total BTEX was between 43% and 45% by volume and toluene was at or above 30.8% by volume, a concentration of toluene that is more than double that typically found in premium gasoline.

The moving front tests showed that premium gasoline did not have sufficient toluene or other BTEX compounds to initiate a moving front. Experiments with toluene in NIST reference fuel and toluene-fortified gasoline showed that a concentration of more than 40% (v/v) toluene was required to initiate a moving front. This is well above the concentration of toluene or total BTEX in premium gasoline. Experiments with mixtures of BTEX compounds in reference fuel showed that the initiation of a moving front in PVC pipe is not dependent on the concentration of a single compound but is dependent on sum of the concentrations (i.e., activities) of the BTEX compounds. A moving front will not form in PVC pipe unless the total BTEX is greater that 40% (v/v).

PVC pipes that become permeated with hydrocarbons should be replaced (if they have not burst due to loss of physical strength) because remediation of the pipe material is not feasible.

Predictive Test –Near Infrared (NIR) Reflectance Spectroscopy

NIR spectroscopy was found to be a useful screening tool for predicting pipe permeation susceptibility and for tracking the permeation process of pure solvents such as benzene and toluene in PVC pipes. NIR spectra of unexposed and exposed pipes and laboratory reference data such as percent weight gain, weight gain per unit length and advance of the moving front were needed to establish a database for a given model of NIR spectrometer. Calibration models developed, however, are limited to the types of PVC pipes within the database and may not be able to predict with certainty for the types of PVC pipes not in the database.

Permeation of Gaskets by Hydrocarbons

Ductile Iron Pipes and Tyton® Gaskets

Using pipe-drum experiments with 4-inch DI joints, permeation was observed for ethylene propylene diene monomer (EPDM) after 35 days of exposure to gasoline, and after 50 days for neoprene (CR) and styrene butadiene rubber (SBR). NBR and FKM were the most resistant to permeation with permeation after 124 days for NBR while very low concentrations of BTEX were detected for FKM without a sudden breakthrough of BTEX, which was probably attributable to leakage, not permeation.

Upon exposure to gasoline-saturated water (about 168 mg/L total BTEX), breakthrough was observed through the SBR gasket after 210 days. Upon exposure to a 50% saturated aqueous solution of gasoline, breakthrough was observed after 240 days. No permeation was observed through SBR gaskets exposed to 20%, and 5% saturated aqueous solutions of gasoline.

No permeation through NBR gaskets was observed for any level of gasoline contamination in water.

Experiments with 4-inch DI pipe joints using SBR Tyton® gaskets exposed to free product premium gasoline for different hydrostatic pressures (0, 20, 40, and 60 psig) indicated that there was no correlation between permeation and hydrostatic pressure in the pipe.

In a service line where periods of stagnation occur, the 5 μg/L MCL for benzene was predicted to be exceeded during an 8-hour stagnation period for SBR gaskets in contact with free product gasoline. For the same conditions, the benzene MCL in the pipe-water was predicted to not be exceeded for DI pipe using NBR gaskets. For SBR Tyton® gaskets exposed to gasoline in groundwater, even groundwater that is saturated with gasoline, the EPA MCL for benzene would probably not be exceeded in the pipe water even for 8 hours of stagnation.

With a minimal average water flow, benzene in the pipe water was predicted to not exceed the MCL for DI pipes with SBR Tyton® gaskets exposed to gasoline, even under the worst conditions of gasoline contamination.

PVC Pipes and Rieber Gaskets

Permeation of gasoline through PVC pipes with SBR and NBR Rieber gaskets was rapid with breakthrough of benzene in 21 days. The MCL for benzene (5 μg/L) was exceeded at breakthrough.

After 8 hours of stagnation in contact with free product gasoline, the concentration of benzene in pipes of diameters up to 24 inches using SBR gaskets was predicted to exceed the MCL. Under the same conditions, the MCL would be exceeded in pipes of diameters up to 10 inches using NBR gaskets. In pipes of diameters ≥12 inches with NBR Rieber gaskets, the concentration of benzene was not predicted to exceed the MCL. Flow estimations indicate that minimal average water flow would be sufficient to dilute the benzene to less than the MCL for both SBR and NBR gaskets when exposed to free product gasoline. Therefore, in typical gasketed PVC pipes with full water flow, the concentration of benzene was predicted to not exceed the MCL due to dilution.

After 9 months of testing, there was no permeation of BTEX through either SBR or NBR gaskets exposed to any level of gasoline contamination in water (including 100%, 75%, 50%, or 10% saturated aqueous solutions of gasoline).

RECOMMENDATIONS

PE Pipe

For practical purposes, there is no level contamination at which PE pipe is resistant to permeation by gasoline, chlorinated solvents, or aqueous solutions of gasoline or chlorinated solvents. Corrective action should be taken whenever there is a known spill of gasoline or organic solvents in the vicinity of a PE service line. Regardless of whether there is a taste or odor complaint, it is most likely that the MCL for benzene will be quickly exceeded in the potable water since the taste and odor threshold level for benzene in potable water is well above the MCL for benzene.

PVC Pipe

PVC pipe is impervious to gasoline. Where gasoline is the only source of hydrocarbon contamination and a choice is to be made between PVC or DI pipe, engineering considerations other than permeation should govern the selection of pipe materials. However, if conditions of stagnation and contact with free product are anticipated, selection of DI pipe with NBR gaskets is preferable, as discussed below.

PVC pipe is impervious to benzene, toluene, and TCE at concentrations in water that are commonly encountered in the field. Based on the laboratory experiments, moving fronts were not observed for less than 60% aqueous solutions of benzene, toluene, and TCE but weight gains of 2% in the sorption test were observed for 60% aqueous solutions of toluene and TCE.

AWWA standards relating to plastic pipes and gasketed pipes include a statement regarding permeation by hydrocarbons directing utilities to consult with manufacturers before selecting pipe or gasket materials to be used in contaminated or potentially contaminated areas. PVC pipe manufacturers commonly refer customers to the Uni-Bell PVC Pipe Association for guidance regarding use of PVC pipes in contaminated areas. The current criteria for compatibility of PVC pipe recommended by the Uni-Bell PVC Pipe Association are activities of 0.25 for benzene and toluene and activities of 0.10 for chlorinated solvents. These criteria were based on sorption experiments involving exposure of thin films of pure PVC polymer powders to solvents in the vapor phase and exposure of PVC sheets pressed from pure powders to the solvents.

The laboratory results reported here indicate that PVC pipes can tolerate a higher activity than that specified by Uni-Bell and that the impact of the individual activities of BTEX compounds is additive. PVC pipe is compatible with benzene, toluene, ethylbenzene, and xylenes (BTEX) in groundwater, provided that the sum of the activities of the individual compounds does not exceed 0.40 and no other swelling solvents are present in significant concentrations. This guidance includes a margin for safety since laboratory results showed that BTEX activities (or TCE activity) must exceed 0.60 activity for a moving front to form and permeation to occur. Activity in this case means the concentration of a compound expressed as a decimal fraction of its maximum solubility in water or vapor. PVC is compatible with TCE in groundwater provided that the activity of TCE does not exceed 0.40 and no other swelling solvents are present in significant concentrations. Mixtures of TCE and BTEX were not studied in this project.

Tyton® Gaskets

SBR Tyton® gaskets may be used (without exceeding the MCL for benzene) in DI water mains under all conditions of gasoline contamination as long as there is at least minimal average flow in the main (at least 0.24 gpm in 4-inch pipe) and conditions of stagnation do not occur. Under conditions of stagnation, as for a service line in direct contact with free product gasoline or soil vapor from a nearby layer of free product, NBR gaskets may be used (without exceeding the MCL for benzene). NBR Tyton® gaskets are the most cost effective choice when a gasket material resistant to gasoline is desired.

The guidance offered in this report is based on the assumption that water is safe to drink if the U.S. EPA maximum contaminant levels (MCLs) are not exceeded for any contaminant. Sufficient information is given in this report for utilities to perform the equivalent calculations for lower contaminant levels. It should be noted that the EPA's maximum contaminant level goal (MCLG) for benzene is 0 μg/L.

Rieber Gaskets

Both SBR and NBR Rieber gaskets may be used (without exceeding the MCL for benzene) in PVC water mains in direct contact with free product gasoline provided there is a minimum average water flow (at least 0.037 gpm in 2-inch pipe). For conditions of 8-hour stagnation in contact with free product gasoline, the MCL was predicted to be exceeded for pipes using SBR Rieber gaskets up to 24 inches in diameter and for NBR Rieber gaskets up to 10 inches in diameter. Either SBR or NBR Rieber gaskets may be used (without exceeding the MCL for benzene) for PVC pipes exposed to gasoline-saturated water.

FUTURE RESEARCH

This study has extended previous research by conducting laboratory experiments for concentrations that were similar to field conditions by using various aqueous solutions of gasoline, benzene, toluene and TCE. In addition, this study has developed a predictive test method (moving front) for PVC pipes and has assessed the utility of NIR spectroscopy.

The moving front test can be easily used to test aged PVC pipes and to study whether aging increases susceptibility. Since it is difficult to obtain aged pipes and their history, a deliberate attempt should be made to age the pipes in the laboratory. This could include burying and exposing pipes to soils of different pHs, for different period of times, and exposure to sunlight/ultraviolet light. These pipes could then be tested using the moving front test and the progression of the moving front would provide information on the susceptibility of the pipe under different aging conditions.

The research has focused on some of the major hydrocarbons encountered in the field such as BTEX in gasoline, and chlorinated solvents such as TCE. Testing should continue by using the predictive test methods and pipe-bottle apparatuses to assess the permeation of similar light aromatic compounds such as ethylbenzene, xylenes, and chlorinated solvents such as tetrachloroethylene (PCE), dichloroethylene (DCE), tetrachloroethane (PCA), and trichloroethane (TCA) to determine whether these chemicals behave in a manner similar to benzene, toluene and TCE. While the impact of on permeation of mixtures of BTEX compounds was studied in this project, future research should investigate permeation by mixtures of chlorinated solvents and by mixtures of chlorinated solvents and BTEX compounds.

Of interest is a further understanding of the pathway of contamination or the permeation process through the Tyton® and Rieber gaskets. The Tyton® and Rieber gaskets were designed to provide a tight joint to prevent water leakage under hydrostatic pressures and may not be optimized in its design to minimize permeation from the outside. Further research should focus on using better visualization or analytical tools to examine exposure areas, the path of travel for the contaminants, and modeling the permeation by using equivalent gasket dimensions. By building a better mathematical model, permeation can be estimated for different materials if the diffusion coefficients are determined using one or more of the predictive test methods.

There is a need for a more definite predictive test method to assess the permeability of gasket materials. The gravimetric sorption test method serves as a rapid method but does not provide sufficient data to provide relative magnitudes of the permeation rates.

CHAPTER 1
INTRODUCTION AND BACKGROUND

OVERVIEW

Over the past three decades, plastic pipes have been used increasingly as a means for the conveyance of drinking water in water distribution systems. In many countries, plastic pipes are the preferred materials. Historically, cast iron and cement or cement-lined pipes have been used but they are now being increasingly replaced by plastic pipes. The ease of installation and handling of plastic pipes is one factor driving the increased use of plastic pipes. Plastic pipes are also highly durable and offer good resistance to the chemicals used in water treatment such as chlorine. Plastic materials commonly used for the distribution of drinking water include polyvinyl chloride (PVC), polyethylene (PE), and polybutylene (PB).

However, in many urban areas, soils have become contaminated as a result of leaking underground storage tanks, chemical spills from industrial areas, improper disposal of used chemicals, and abandoned waste sites. These pollutants from leaking storage tanks and contaminated soils can and have posed serious threats to the longevity and structural integrity of plastic (PE and PVC) pipes and elastomeric gaskets which, in turn, can affect the water quality in the distribution system. Many state agencies and municipalities have policies prohibiting use of plastic pipes in hydrocarbon-contaminated subsurface environments. Although there are research studies and case studies documenting the permeation of organic compounds through plastic pipes and elastomeric gaskets, and the impact on water quality, there is still a lack of understanding on the performance of PE and PVC pipe materials, and elastomeric gaskets in hydrocarbon-contaminated soils commonly encountered under field conditions.

OBJECTIVES

The overall objective of this project was to study the impact of petroleum-based hydrocarbons on PE and PVC pipes, and elastomeric gaskets. This project advanced the water industry's understanding of the parameters that influence permeation of plastic pipes under conditions of soil contamination that are typically encountered in the field. This increased understanding provides guidance to utilities regarding the conditions under which plastic pipes can be used in contaminated areas. The specific objectives of this project were:

i. Conduct a survey of water utilities to obtain information on the types of pipe material and gasket materials used for mains and service connections. Other information included hydrocarbon permeation incidents and successful use of plastic pipes (PE/PVC) or gasketed pipes in contaminated areas.

ii. Conduct hydrocarbon permeation experiments using the pipe-drum or pipe-bottle apparatus for PE/PVC pipes under simulated environmental conditions. Environmental conditions included saturated and unsaturated soil conditions, different levels of hydrocarbon contamination and three different soil types (for soil types, tests were conducted only for PE pipes).

iii. Develop predictive laboratory tests that may predict the performance of plastic pipe materials in soils contaminated with petroleum-based hydrocarbons. Predictive laboratory tests included the equilibrium sorption test, the moving front test (visualization test) and near infrared (NIR) reflectance spectroscopy.

iv. Conduct pipe-drum experiments under simulated environmental conditions to assess the performance and failure of gasket materials exposed to hydrocarbon contamination. Materials tested included styrene-butadiene-rubber (SBR), neoprene (CR), ethylene-propylene-diene-monomer (EPDM), nitrile rubber (NBR), and fluoroelastomer rubber (FKM). Equilibrium sorption tests for gasket materials were studied as well as the impact of hydrostatic pressure on hydrocarbon permeation through Tyton® gaskets used with ductile iron pipe.

SIGNIFICANCE OF THE PROJECT

The project advances the water industry's understanding of the parameters that influence permeation of plastic pipes and elastomeric gaskets under conditions of soil contamination that are typically encountered in the field. This increased understanding enables the development of guidelines for utilities with respect to the contamination concentration levels, soil conditions (soil types, saturated or unsaturated), and pipe materials that can be used. The project highlights the boundary limits and environmental conditions which will result in permeation.

A survey of water utilities on permeation incidents involving plastic pipes and elastomeric gaskets provided information on utilities' experiences. This survey complements an earlier survey conducted about twenty years ago. The survey identified the nature and extent of hydrocarbon contamination commonly encountered in the field, types of plastic pipes commonly used, and susceptibility of pipes to permeation, and provided information for the design of laboratory experiments to study permeation under simulated environmental conditions.

Laboratory experiments conducted under realistic environmental conditions provided information on the performance of PE/PVC pipes and gaskets in the presence of hydrocarbons. Predictive laboratory tests were also developed that can provide a rapid means of comparing the relative susceptibility to permeation of different plastic pipes and elastomeric gaskets.

REVIEW OF PREVIOUS STUDIES

Pipes and Contaminants

Common plastic pipe materials, listed in decreasing order of susceptibility to permeation, are: polybutylene (PB) > polyethylene (PE) > polyvinyl chloride (PVC). In a national survey done for AwwaRF (Thompson and Jenkins 1987), 43% of reported permeation incidents in the U.S. involved PB pipe, 39% involved PE pipe, and 15% involved PVC. Acrylonitrile-butadiene-styrene (ABS), asbestos-cement, and gasket materials were each involved in 1% of the permeation incidents reported in the survey. Although gasket materials are relatively permeable, the low rate of permeation incidents involving gaskets was attributed to their use primarily in water mains with high flow rates and little stagnation. Most of the reported incidences of foul odor due to permeation by organic compounds involved PB and PE service connections that were subject to prolonged periods of stagnation. Most of the incidents in the survey involved heavy contamination in the soil surrounding the pipe.

Nearly all permeation incidents in the U.S. (from early 1970s to 1985) were related to petroleum products, mainly gasoline spills (Thompson and Jenkins 1987, Holsen et al. 1991a). The monoaromatic compounds in gasoline, benzene, toluene, ethylbenzene and o-, m-, and p-xylene (BTEX), permeate PB and PE pipes readily (much more readily than the aliphatic compounds) and are the compounds of concern in permeation incidents. A small number of incidents involved chlorinated solvents such as trichloroethylene (TCE), which also permeate PB and PE readily. While the incidence was small, chlorinated solvents are a concern because some of the solvents are carcinogens, are persistent in the environment, and are present in soil at most contaminated sites. BTEX and chlorinated solvents also soften PVC pipe by rubberizing its otherwise rigid, glassy polymer structure, thereby causing it to be susceptible to permeation in a manner similar to PB and PE.

Gaskets

Styrene-butadiene rubber (SBR) is the most common synthetic rubber material used in pipe gaskets while nitrile (NBR), ethylene propylene diene monomer (EPDM), neoprene (CR), and fluoroelastomer rubber (FKM) are also available and used for the specific environmental applications. Olson et al. (1987) studied the permeation of toluene, hexane and 1,1,1-trichloroethane (pure solvents) through gasketed pipes of PVC, asbestos cement and ductile iron pipes and found that all three gasketed pipe systems became permeated. Seo and Kim (1998) conducted permeation experiments using cast iron pipes with SBR gaskets in the presence of gasoline and toluene (pure solvents) and found that the breakthrough times were approximately 100 days. Glaza and Park (1992) conducted permeation experiments with ductile iron pipes using SBR, NBR and a special NBR gasket in the presence of gasoline. They found that NBR was more resistant to gasoline permeation than SBR and the order of breakthrough for different compounds was benzene, toluene, ethylbenzene, m-xylene and o+p-xylene. The results of the previous studies indicate that gaskets are likely to be susceptible to permeation by various organic compounds. However, the permeation studies conducted were for extreme conditions using pure solvents and gasoline. In fact, there are very few known studies for concentration levels typically found at contaminated sites and groundwater.

Physicochemical Theory of Permeation

Different diffusion mechanisms occur in PE and PVC because of fundamental differences in their polymer structures. PVC is an amorphous glassy polymer with very limited flexibility of the polymer chains. PE is a partially crystalline, rubbery polymer having amorphous areas with high polymer chain end mobility. Small organic molecules permeate PVC through the small free volumes between the relatively immobile polymer chains, whereas permeation of PE occurs through the amorphous areas having relatively more mobile polymer chains. This structural difference accounts for the dissimilarity in permeation mechanisms between PE and PVC. PB behaves similarly to PE (Holsen et al. 1991a), except that it is more permeable.

Permeation of plastic water pipes and gasket materials by hydrocarbons occurs by a three-step process consisting of: (i) partitioning of the contaminant between the ground water or soil-gas and the outer surface of the polymer; (ii) diffusion of the contaminant through the polymer; and (iii) partitioning of the contaminant between the inner surface of the polymer and water in the pipe. A concise description of this process is given by Vonk and Veenendaal (1984) and a detailed treatment of the mathematics of diffusion is given by Crank and Park (1968). Both partitioning steps proceed according to Henry's law, wherein the concentration of contaminant in the receiving phase is equal to the concentration in the transmitting phase multiplied by a constant (i.e., Henry's Law constant), which is characteristic of the contaminant and the two phases. The diffusion of the contaminant through the polymer results in a concentration gradient within the thickness of the polymer that can be described by the laws of Fick, provided that the contaminant is inert to the polymer, i.e., there are no solvency or swelling effects.

Fick (1855) observed that both diffusion and heat transfer by conduction are due to random molecular motion, and that Fourier's (1822) equation of heat conduction could be adopted to describe diffusion. The diffusion equation is known as Fick's First Law and is based on the hypothesis that the rate of diffusion is dependent on the concentration gradient across the thickness of a material. Barrer (1951) combined the expressions of Henry's and Fick's laws to derive an expression for the total mass of contaminant permeating a polymer as a function of time, $Q(t)$ (g/m^2), under the simplifying assumptions: (i) that the outer concentration of contaminant remains constant; (ii) the initial concentration of the contaminant in the polymer is zero; and (iii) the inner concentration of the contaminant is kept at zero (similar to a constantly flowing water main).

For plastic water pipes, the rate of permeation ($dQ(t)/dt$) is directly proportional to the concentration of the contaminant surrounding the pipe and inversely proportional to the pipe wall thickness. Accordingly, in a contaminated soil, doubling the pipe wall thickness will quadruple the time required for contamination to reach detectable levels, and will reduce the steady-state rate of permeation by one half. These statements apply for any contaminant-polymer pair for which the contaminant is otherwise inert to the polymer. It is conceivable; therefore, that using pipe of greater wall thickness may reduce the frequency of detectable permeation incidents.

The diffusion coefficient, D, is markedly dependent on the concentration of the contaminant diffusing through the polymer. This effect is attributed to increased mobilities of polymer segments resulting from increases in the average free volume in the polymer caused by the presence of the diffusing contaminant (Crank and Park 1968). The diffusion coefficient is dependent on the volume and shape of the contaminant molecule. Berens and Hopfenberg (1982) found that for molecules which are roughly spherical in shape such as the noble gases, methane, carbon tetrachloride, and neopentane, the diffusion coefficient in PVC increases exponentially with a decrease in molecular diameter. For elongated molecules such as butane and larger n-alkanes, the diffusion coefficient is larger than would be predicted for a spherical molecule of equivalent volume, but not as large as would be predicted by the minimum diameter of its shape. This suggests that anisometric (not isometric) molecules pass through the polymer structure of PVC by moving along their long axes. Larger molecules, such as p-nonylphenol and tetrachlorodioxin, are predicted to take hundreds to millions of years to diffuse through PVC, based on diffusion coefficients calculated from their mean molecular diameters (Berens 1997). Other researchers have also found that diffusion coefficients decrease strongly with increasing molecular diameter in rubber (Guo et al. 1992) and other materials (Menting et al. 1970, Xiao and Wei 1992, and Satterfield and Cheng 1971).

Examples of diffusion coefficients (cm^2/s) at low contaminant activity reported in the literature include:

Low density polyethylene pipe (LDPE):	Toluene – 0.39×10^{-8} TCE - 0.6×10^{-8}
High hensity polyethylene pipe (HDPE):	TCE - 0.2×10^{-8}
Polyvinyl chloride pipe(PVC):	Toluene - 0.9×10^{-24}
Styrenebutadiene rubber gaskets (SBR):	Toluene – 1.7×10^{-8}

Although Fickian diffusion occurs in PVC, the diffusion coefficients calculated from sorption experiments on PVC powders and films suggest that PVC should resist permeation for centuries (Berens 1985). Prior to the AwwaRF sponsored survey of 1987 (Thompson and Jenkins 1987), PVC pipe had been thought to be essentially immune to permeation by environmental contaminants (Holsen et al. 1991a).

Permeation by means of softening or swelling of the polymer structure is called Case II diffusion. Fickian diffusion proceeds so slowly in PVC that Case II diffusion predominates if an organic contaminant with a solvency effect on PVC is present in sufficient concentration to cause swelling. Case II diffusion results when PVC is progressively converted from a glassy plastic with a low Fickian diffusion coefficient to a rubbery plastic with a high diffusion coefficient. The zone of swelling progresses through the pipe wall as a "moving-front" (Vonk and Veenedaal 1984). Rubberized PVC shares similar characteristics with PE of having a relatively high permeability to organic compounds, owing to increased mobility of the polymer chains. This increased mobility is thought to be caused by disruption of inter-chain hydrogen bonds to chlorine atoms by the contaminant molecules dissolved in the polymer structure. Case II diffusion may also result in a loss of structural integrity in PVC pipe.

Unlike Fickian diffusion in which the rate of permeation increases to a constant rate once steady-state diffusion has been reached within the polymer, no detectable permeation occurs in Case II diffusion until the moving front of rubberization reaches the inside wall of the pipe. At that point, the contaminant suddenly breaks through at its maximum constant rate of permeation.

Activity and Permeation

Predictions of the susceptibility of pipe material to permeation by organic compounds are usually based on the activity of a compound rather than the concentration (Vonk 1986, Berens 1985). Berens (1985) used the pure liquid as the reference state and assigned the pure liquid an activity of one. Consequently, the saturated vapor in equilibrium with the pure liquid was also assigned an activity of one. Vonk (1986) defined activity in the context of permeation theory by using aqueous solubility as follows:

$$\alpha_i = C_w/C_{w,m} \quad \text{or} \quad \alpha_i = C_v/C_{v,m} \tag{1.1}$$

where: α_i = activity ($0 \leq \alpha_i \leq 1$) (not to be confused with chemical activity coefficient)
C_w = concentration in water (mg/L)
$C_{w,m}$ = maximal (saturated) solubility in water (mg/L)
C_v = concentration in the vapor phase (mg/L)
$C_{v,m}$ = maximal (saturated) concentration in the vapor phase (mg/L).

Based on Vonk's definition, an activity of one is when the concentration of a compound is equal to it maximum solubility in water.

The activity of one for both Berens' and Vonks' reference states are the same since the vapor pressure of a pure solvent will be the same as the vapor pressure for a saturated solution of the solvent. The uncertainty, however, is for activities that are less than one. Vonk indicated that for hydrocarbons that soften PVC readily such as chlorinated hydrocarbons, ketones, nitrobenzenes, and some anilines, activities of ≤ 0.1 (0.1 times the maximal concentration in water) result in pure Fickian diffusion and no softening of PVC occurs. For compounds which soften PVC somewhat less readily, such as BTEX compounds, the activities indicated for Fickian diffusion by Behrens (1985) are ≤ 0.25. At or below these activities, it would take thousand of years before contamination would become detectable in PVC pipe (Vonk 1986). The work of Berens (1985) showed that at ambient temperature, PVC material will be softened only by strong solvents at activities (% by volume of the solvent) > 0.5. For activities up to 0.25, transport of organic solvents in PVC pipe follows Fickian diffusion and therefore PVC pipe is an effective barrier for these activities. As seen from the two studies above, the absolute values of the activities that would result in permeation of PVC pipes may be similar but the reference states of the activity used were different.

Effect of Soil Type

Vonk and co-workers in the Netherlands found that sorption of organics on soil particles in the unsaturated zone can reduce the effective availability for permeation of those compounds to a fraction of what might otherwise be inferred from a chemical analysis of the soil (Vonk, 1986). Holsen et al. (1991b), in experiments with PB pipes, found that pipes buried in high organic soils will be permeated more slowly than those buried in low organic soils. Since hydrocarbons are likely to be associated with the organic matter in soil, a soil with high organic content can effectively lower the partial pressure of a contaminant in the soil. Since permeation in the unsaturated zone takes place in the vapor phase, reducing the partial pressure reduces the activity of a contaminant. Accordingly, for a given contaminant concentration, plastic pipes in high organic soils are less susceptible to permeation by hydrocarbons than those in low organic soils.

Holsen et al. (1991b) recommended that plastic pipes not be backfilled with sand but rather with the native soil because the organic content could provide some protection from permeation. Holsen further stated that soils with high organic content cannot completely protect plastic pipes from permeation but can decrease the number of permeation incidents.

Additives

PE and PVC are produced as pure powders which must then be extruded to form pipes. Before pipes can be extruded, various additives such as stabilizers and lubricants are mixed with the pure powders to obtain the desired properties of the pipe and to assure successful extrusion. Stabilizers (antioxidants) are added to prevent thermal decomposition during extrusion. Lubricants are added to prevent the material from sticking to the extruder. A great variety of additives are available for use in polyolefin plastics (including PE and PB pipes). For example, in the early 1990's, patents for antioxidants for polyolefins were being recorded at a rate exceeding 380 per year worldwide (Vasile and Seymour, 1993). Stabilizers for older PVC pipes included organic and inorganic compounds of lead, cadmium and tin, but recent improvements in extrusion machinery design and process control has enabled use of complexes of barium, calcium, magnesium, strontium and zinc as stabilizers. Lubricants for PVC include stearates and waxes (Matthews 1996).

There is virtually no information in the literature about the influence of additives on permeation by hydrocarbons. All of the research has been done on pure polymer powders, on sheets pressed from the pure powders, or on extruded pipes for which the compounding is unspecified except for the main polymer composition. Berens (1985) reported that the equilibrium sorption and sorption rates of acetone, toluene and TCE liquids in finished PVC pipe and sheets pressed from a commercial pipe powder compound (additives not specified) were equivalent to the results obtained using pure PVC.

SUMMARY

In a previous utility survey, Thompson and Jenkins (1987) reported that about 82% of all documented permeation incidents involved PE or PB pipe. PE pipe is susceptible to permeation by organics at low levels of contamination. PVC pipe, for practical purposes, is resistant to permeation by organics except in conditions of gross environmental contamination but the gasket materials are highly permeable. Based on current literature, gasketed water mains (using SBR and EPDM), regardless of the pipe material used, are susceptible to permeation through the gaskets. Thompson and Jenkins (1987) speculated that the extremely low number of permeation incidents attributed to gaskets was due to high flow rates and lack of water stagnation in the pipe.

Many of the laboratory studies conducted used pure solvents or gasoline which typically represents extreme or gross contamination conditions. Although there is a need to understand the risk of permeation at different concentration levels, there are very few studies available to assist in this assessment.

CHAPTER 2
WATER UTILITY EXPERIENCES AND CASE HISTORIES

INTRODUCTION

The water industry was previously surveyed by Thompson and Jenkins (1987) to obtain information about the use of plastic pipes in distribution systems. Their approach involved a postal survey which sought information about a utility's use of plastic pipe and whether or nor permeation incidents had been experienced by the utility. Case histories for the permeation incidents were developed by telephone contacts with the respondents at a later time.

Since the survey described here focused specifically on permeation, detailed information about permeation incidents was requested in the postal survey. The information sought would generally require respondents to search their utility's records and, therefore, obtaining such information by direct telephoning was not considered feasible. Seeking detailed information about permeation incidents in the postal survey was considered a more effective use of time for the respondents and the project team and more likely generate useful information for the project.

The survey done by Thompson and Jenkins (1987) was used as a starting point to develop the survey instrument. The services of a graphic designer were used to develop a survey instrument that was visually appealing and provided a simple layout to obtain complex information. The final version of the survey instrument reflected inputs by the project's utility partners and the AwwaRF Project Advisory Committee.

SURVEY INSTRUMENT

The survey instrument consisted of four pages in landscape format (8.5 ins. x 11 ins.) offset printed on a single-folded 11 ins. x 17 ins. sheet of white paper. Page 1 contained a cover letter, as well as a space to enter contact information for the respondent (Figure 2.1). Page 2 requested information concerning plastic pipe and pipe gasket materials used in mains and summary information concerning permeation incidents involving mains (Figure 2.2). Page 3 requested the equivalent information for service connections (SCs) (Figure 2.3). Page 4 requested detailed case history information for permeation incidents (Figure 2.4). Space was provided to describe three incidents, with a notation that the page could be copied for additional incidents. An insert page (Figure 2.5) was included explaining the acronyms and abbreviations used in the survey instrument. A cover letter from the AwwaRF Project Manager (Figure 2.6) requesting cooperation by the utilities accompanied the survey, along with a postage-paid reply envelope. These materials were sent in a 9 ins. x 12 ins. white envelope bearing a return address printed in Iowa State University's visual identity font.

WATER UTILITY MAILING LIST

Significant difficulties were encountered with the American Water Works Association (AWWA) mailing list. The AWWA mailing list consisted of roughly 55,000 addresses and was administered for public rental by Edith Roman Associates of Pearl River, New York. The data base was structured in such a way that there was no possibility to select "member utilities".

WATER UTILITY SURVEY Impact of Petroleum-based Hydrocarbons on PE/PVC Pipes and Pipe Gaskets

Dear AWWA Member Utility:

Iowa State University's Environmental Engineering Research Laboratory is conducting a project funded by the Awwa Research Foundation (AwwaRF) to study the impact of petroleum-based hydrocarbons on polyethylene and polyvinyl chloride pipes and on elastomeric gaskets. This project will advance the water industry's understanding of the parameters that influence permeation of plastic pipes under conditions of soil contamination typically encountered in the field. A description of the project can be found at www.awwarf.org/research/TopicsAndProjects/projectSnapshot.aspx?pn=2946.

The enclosed survey is a critical first step in this project. This survey will build upon the one taken almost 20 years ago by AwwaRF. Completing the survey will help us produce the best possible guidance on this problem for the water industry. A report on the results of the study will be published by AwwaRF.

Please provide a description of your water mains and service connections on pages 2 and 3. On page 4, please provide brief descriptions of permeation incidents. Please provide as much information as possible.

Please return the survey using the enclosed postage-paid envelope by ______.

Thank you for your cooperation. Please contact us if you have any questions.

Sincerely,

S. K. Ong, Ph.D., P.E.
Associate Professor
Environmental Engineering
Phone: 515.294.3927
E-mail: skong@iastate.edu

Jim Gaunt
Manager
Environmental Engineering Research Lab
Phone: 515.294.8768
E-mail: jagaunt@iastate.edu

Awwa Research Foundation
Advancing the Science of Water®

IOWA STATE UNIVERSITY

Primary Utility Contact

Utility Name ______________________________

Contact Name ______________________________

Title ______________________________

Address ______________________________

____

____

Telephone ______________________________

E-mail ______________________________

Return survey to:
Dr. S. K. Ong
Iowa State University
Dept. of Civil, Construction, and Environmental Engineering
486 Town Engineering Building
Ames, Iowa 50011
Phone: 515.294.3927
Fax: 515.294.8216

Iowa State University does not discriminate on the basis of race, color, age, religion, national origin, sexual orientation, sex, marital status, disability, or status as a U.S. Vietnam Era Veteran. Any persons having inquiries concerning this may contact the Director of Equal Opportunity and Diversity, 3680 Beardshear Hall, 515-294-7612. 04554

Figure 2.1 Water Utility Survey, page 1

MAINS *(Please provide as much information as possible.)*

Total miles of mains in system?________

Miles of mains you consider at risk of permeation by petroleum-based hydrocarbons? ________________

Do you have a protocol regarding the installation or replacement of plastic pipes or pipe gaskets in hydrocarbon-contaminated areas? ○ Yes ○ No

If yes, briefly describe.

____

____

____

____

____

In the table at right:

1. Enter the miles of each type of main.
2. Enter the dates mains were installed.

3 4 5 Circle all standards specified for pipes, gaskets, and gasket class designations *(see insert for full names of standards and ASTM class designations).*

6. Enter number of permeation incidents for each type of main.
7. Enter number of those incidents that resulted in customer complaints.
8. Enter number of known successful uses of plastic mains or gasketed joints in contaminated areas *("successful use" means pipe in contaminated soil without resulting in customer complaints or exceeding EPA MCLs in water).*

	Polyvinyl Chloride (PVC)	Polyethylene (PE)	Ductile Iron (DI)	Cast Iron/Steel (CI/S)	Asbestos Cement (AC)	Concrete	Other ______
(1) Miles							
(2) Period of Use (year)	to	to					
(3) Pipe Specifications	D1785 D2241 D2672 D2740 C900 C905 C909 Other______	D2104 D2239 D2447 D3035 C906 Other______					
(4) Gasket Materials ASTM Class Designations	BR SBR EPM EPDM NBR CR Other______ No gasket	BR SBR EPM EPDM NBR CR Other______ No gasket	BR SBR EPM EPDM NBR CR Other______ No gasket	BR SBR EPM EPDM NBR CR Other______ No gasket	BR SBR EPM EPDM NBR CR Other______ No gasket	BR SBR EPM EPDM NBR CR Other______ No gasket	BR SBR EPM EPDM NBR CR Other______ No gasket
(5) Gasket/Joint Specifications	F477 F913 Other______	F477 F913 Other______	F477 C115 F913 Other______	F477 F913 Other______	F477 F913 Other______	F477 C301 F913 Other______	F477 F913 Other______
(6) No. of Permeation Incidents							
(7) No. of Resulting Customer Complaints							
(8) No. of Successful Uses in Known Contaminated Areas							

Figure 2.2 Water Utility Survey, page 2

SERVICE CONNECTIONS *(Please provide as much information as possible.)*

Number of service connections (of all types) in system? ________

Do your local codes allow plastic pipe for service connections? ◯ Yes ◯ No

Number of services you consider to be at risk of permeation by petroleum-based hydrocarbons? ________

In the table at right:

(1) Enter number of each type of service connection.
(2) Enter dates those types of materials were installed.
(3)(4)(5) Circle all standards specified for pipes, gaskets, and gasket class designations *(see insert for full names of standards and ASTM class designations).*
(6) Enter number of permeation incidents for each type of service connection.
(7) Enter number of those incidents that resulted in customer complaints.
(8) Enter number of those incidents that were related to termiticides.
(9) Enter number of known cases of successful use of plastic service connections in contaminated areas *("successful use" means pipe in contaminated soil without resulting in customer complaints or exceeding EPA MCLs in water).*

	Polyvinyl Chloride (PVC)	Polyethylene (PE)	Polybutylene (PB)	Copper	Ductile Iron (DI)	Cast Iron/ Steel (CI/S)	Lead	Other ____
(1) Number of Connections								
(2) Period of Use (year)	to	to						
(3) Pipe Specifications	D1785 D2241 D2672 D2740 C900 C905 C909 Other____	D2104 D2239 D2447 D3035 C906 C901 C903 Other____						
(4) Gasket Materials ASTM Class Designations	BR SBR EPM EPDM NBR CR Other____ No gasket	BR SBR EPM EPDM NBR CR Other____ No gasket	BR SBR EPM EPDM NBR CR Other____ No gasket		BR SBR EPM EPDM NBR CR Other____ No gasket	BR SBR EPM EPDM NBR CR Other____ No gasket		BR SBR EPM EPDM NBR CR Other____ No gasket
(5) Gasket/Joint Specifications	F477 F913 Other____	F477 F913 Other____	F477 F913 Other____		F477 F913 Other____	F477 F913 Other____		F477 F913 Other____
(6) No. of Permeation Incidents								
(7) No. of Resulting Customer Complaints								
(8) No. of Incidents Related to Termiticides								
(9) No. of Successful Uses in Known Contaminated Areas								

Figure 2.3 Water Utility Survey, page 3

PERMEATION INCIDENTS *(copy form for additional incidents)*

A permeation incident is hydrocarbon contamination that results in customer complaints regarding taste, odor, or illness, or that produces test results for BTEX, TCE, etc., that exceed EPA MCLs.

	INCIDENT 1 Date ________	INCIDENT 2 Date ________	INCIDENT 3 Date ________
Successful Use*	Yes No	Yes No	Yes No
Type	Main Service Connection	Main Service Connection	Main Service Connection
Customer Complaint	Odor Taste Illness Other ________ None	Odor Taste Illness Other ________ None	Odor Taste Illness Other ________ None
Contamination	Gasoline Chlorinated Solvents Diesel Fuel Oil Other ________	Gasoline Chlorinated Solvents Diesel Fuel Oil Other ________	Gasoline Chlorinated Solvents Diesel Fuel Oil Other ________
Source of Contamination	LUST** Dry Cleaning Superfund Disposal Site Spill Other ________	LUST** Dry Cleaning Superfund Disposal Site Spill Other ________	LUST** Dry Cleaning Superfund Disposal Site Spill Other ________
Exposure Time Prior to Failure/Complaint	________ months ________ days	________ months ________ days	________ months ________ days
Concentrations *Example: BTEX, TPH, TCE in mg/L, mg/kg, or ppm*	**Soil** ________ = ________ mg/kg **Ground water** ________ = ________ mg/L **Potable water** ________ = ________ mg/L	**Soil** ________ = ________ mg/kg **Ground water** ________ = ________ mg/L **Potable water** ________ = ________ mg/L	**Soil** ________ = ________ mg/kg **Ground water** ________ = ________ mg/L **Potable water** ________ = ________ mg/L
Soil Type	Sand Loam Silt Clay Not Sure	Sand Loam Silt Clay Not Sure	Sand Loam Silt Clay Not Sure
Organic Content (%)	High (> 5%) Medium (1–5%) Low (< 1%) Not Sure	High (> 5%) Medium (1–5%) Low (< 1%) Not Sure	High (> 5%) Medium (1–5%) Low (< 1%) Not Sure
Type of Pipe	PVC PE PB Other ________	PVC PE PB Other ________	PVC PE PB Other ________
Diameter (inches/mm)/ **Schedule**	inches/mm | Sch 40 Sch 80 Sch 120 Other ________	inches/mm | Sch 40 Sch 80 Sch 120 Other ________	inches/mm | Sch 40 Sch 80 Sch 120 Other ________
Pipe Specification (see insert)			
Gasket – ASTM Class (see insert)	BR SBR EPM EPDM NBR CR No Gasket Other ________	BR SBR EPM EPDM NBR CR No Gasket Other ________	BR SBR EPM EPDM NBR CR No Gasket Other ________
Type of Bedding	Sand Native Soil Other ________ None	Sand Native Soil Other ________ None	Sand Native Soil Other ________ None
Pipe/Gasket Failure	Pipe Gasket	Pipe Gasket	Pipe Gasket
Describe Failure			
Describe Corrective Action and Outcome			

* *"Successful use" means pipe in contaminated soil without resulting in customer complaints or exceeding EPA MCLs in water.*
***Leaking Underground Storage Tank*

Figure 2.4 Water Utility Survey, page 4

Standard Specifications for PE/PVC Pipes and Gasket

PIPES

PVC

ASTM D 1785........ Poly(Vinyl Chloride) (PVC) Plastic Pipe Schedules 40, 80, and 120
ASTM D 2241........ Poly(Vinyl Chloride) (PVC) Plastic Pipe (SDR-PR)
ASTM D 2672........ Bell-End Poly(Vinyl Chloride) (PVC) Pipe
ASTM D 2740........ Poly(Vinyl Chloride) (PVC) Plastic Tubing
AWWA C900.......... Poly(Vinyl Chloride) (PVC) Pressure Pipe and Fabricated Fittings, 4 in.–12 in. (100 mm–300 mm), for Water Dist.
AWWA C905.......... Poly(Vinyl Chloride) (PVC) Pressure Pipe and Fabricated Fittings, 14 in.–48 in. (350 mm–1,200 mm)
AWWA C909.......... Molecularly Oriented Polyvinyl Chloride (PVCO) Pressure Pipe, 4 in.–12 in. (100 mm–300 mm), for Water Dist.

PE

ASTM D 2104........ Polyethylene (PE) Plastic Pipe, Schedule 40
ASTM D 2239........ Polyethylene (PE) Plastic Pipe (SDR-PR)
ASTM D 2447........ Polyethylene (PE) Plastic Pipe, Schedules 40 and 80, Based on Outside Diameter
ASTM D 2737........ Polyethylene (PE) Plastic Tubing
ASTM D 3035........ Polyethylene (PE) Plastic Pipe (SDR-PR) Based on Controlled Outside Diameter
AWWA C901.......... Polyethylene (PE) Pressure Pipe, Tubing, and Fittings, ½ in. (13 mm) through 3 in. (76 mm), for Water Service
AWWA C903.......... Polyethylene-Aluminum-Polyethylene and Crosslinked Polyethylene Composite Pressure Pipes
AWWA C906.......... Polyethylene (PE) Pressure Pipe and Fittings, 4 in. (100 mm) th. 63 in. (1.575 mm), for Water Dist. and Trans.

Other Pipe

AWWA C115.......... ANSI Standard for Rubber-Gasket Joints for Ductile-Iron Pressure Pipe and Fittings
AWWA C301.......... Prestressed Concrete Pressure Pipe, Steel-Cylinder Type

Certification for Use with Potable Water

(These are generally referenced in the above standards for pipes to be used for potable water)

NSF 14.................. ANSI/NSF International Standard for Plastics – Plastics piping system components and related materials
NSF 61.................. NSF/ANSI Standard for Drinking Water Additives – Drinking water system components – Health effects

---

ELASTOMERIC GASKET MATERIALS FOR JOINING PLASTIC PIPE

Specifications

ASTM F 477 .... Elastomeric Seals (Gaskets) for Joining Plastic Pipe
ASTM F 913 .... Thermoplastic Elastomeric Seals (Gaskets) for Joining Plastic Pipe

ASTM Classification

Non-Oil Resistant

BR .... butadiene rubber
SBR........................ styrene-butadiene rubber
EPM .... ethylene-propylene copolymer
EPDM..................... ethylene-propylene-diene terpolymer

Oil Resistant

NBR .... acrylonitrile-butadiene rubber
CR .... chloroprene rubber (neoprene)

Figure 2.5 Water Utility Survey Insert

Awwa
Research
Foundation
Advancing the Science of Water®

6666 West Quincy Avenue
Denver, CO 80235-3098 USA
P 303.347.6100
F 303.730.0851
www.awwarf.org
email: info@awwarf.org

Sponsors Research
Develops Knowledge
Promotes Collaboration

July 7, 2004

Dear Utility Manager:

The Awwa Research Foundation is sponsoring a research project on the impact of petroleum-based hydrocarbons on polyethylene and polyvinyl chloride pipes and pipe gaskets. The project will be conducted by Dr. S. K. Ong and Mr. Jim Gaunt of Iowa State University, Ames, Iowa. This project will develop useful information and advance our understanding of the parameters that influence the permeation of petroleum-based hydrocarbons into plastic pipes under conditions commonly encountered in the field.

A critical step in this project is a survey of utilities on this issue. Since this is a vital part of the research project, we would appreciate your cooperation in completing the survey and returning it to Dr. S. K. Ong.

If you have any questions about the survey or the project, please direct your questions to Dr. S. K. Ong at (515) 294 3927 or skong@iastate.edu. I would be pleased to answer any questions about the work at the Research Foundation. You can reach me at (303) 347 6114.

Thank you for your cooperation.

Sincerely yours,

Jian Zhang, Ph.D., P.E.
Project Manager

Figure 2.6 Water Utility Survey, Letter from AwwaRF Project Manager

A close examination of the rented database revealed apparent duplication of utilities, the absence of some larger utilities, and addresses of inappropriate personnel (i.e., heavy equipment operators). These problems stemmed, at least in part, from the fact that the AWWA mailing list contained the addresses where members receive mail from AWWA. Some of these addresses were home addresses, and were not necessarily in the same cities as their water utility employers. Employees of the same water utility sometimes stated their company name differently or not at all. Any variation in the "company" field was interpreted as a different company. Finally, the "one per company" selection did not necessarily yield the most appropriate person to receive the survey. These factors served to seriously degrade the usefulness of the rented database for surveying AWWA member utilities. The utility list from the AWWA's WaterStats database was also used. However, the apparent absence of many utilities that appear in AWWA's WaterStats database was also troubling.

The rented database was heavily edited to eliminate duplication and to assure delivery to the most appropriate recipients. Utilities appearing in WaterStats were merged with the edited database if a mailing address appeared in their website. "Water Utility Director" was used as the addressee if the director's name could not be determined from the website.

POSTAL SURVEY

The mailing list which was finally used to mail the surveys contained 2,997 U.S. addresses and 97 Canadian addresses. Surveys were mailed to U.S. addresses by bulk mail on July 13, 2004 and to Canadian addresses by EPCOR, a Canadian utility partner, from Edmonton, Alberta by regular mail on July 30, 2004. Replies were requested by September 10, 2004. Post card reminders (Figure 2.7) were sent to U.S. addresses on September 9 and to Canadian addresses on October 5.

The postal survey resulted in a 2.5% response rate with significant gaps in geographical distribution.

TELEPHONE SURVEY

Because of the poor response to the postal survey, a telephone survey was conducted from September 2004 to March 2005 with the goal of obtaining responses from at least 2 large utilities in every U.S. state. The survey procedure consisted of using contact information on a utility's web page to identify an appropriate contact person, explaining the project over the phone, and securing permission to send a survey packet to the contact. Officials of 159 water utilities gave permission to send surveys, resulting in a 47% return rate.

SURVEY RESPONSE

Survey responses were received from 151 water utilities in 50 U.S. states, three Canadian provinces, and the District of Columbia. The goal of having a response from two large utilities in each U.S. state was met for 39 states. For 10 states, responses were received from one large utility; for 2 states, no large utility responded.

WATER UTILITY SURVEY Impact of Petroleum-based Hydrocarbons on PE/PVC Pipes and Pipe Gaskets

Awwa Research Foundation

Advancing the Science of Water®

IOWA STATE UNIVERSITY

Dear AWWA Member or Utility Manager:

We recently sent you a survey seeking information on your water utility's experience with permeation of plastic pipes and pipe gaskets by petroleum-based hydrocarbons. This project is funded by the Awwa Research Foundation and will advance the water industry's understanding of the parameters that influence the permeation of plastic pipes under conditions of soil contamination that are typically encountered in the field. Your response will help us provide the best possible guidance to the water industry.

If you have already returned your completed survey, thank you very much!

If you are not the appropriate person to complete the survey, we would greatly appreciate your forwarding it to the appropriate person.

If you need another copy of the survey or have any questions, please feel free to contact me at 515-294-8768 or jagaunt@iastate.edu.

Thank you,

Jim Gaunt, Manager

Figure 2.7 Water Utility Survey, Reminder Postcard

The location and relative size of the responding utilities are presented in Figure 2.8 and Figure 2.9. Because a water utility's risk of a permeation incident depends on both the miles of mains and the number of SCs, the relative size of utilities was classified in Figures 2.8 and 2.9 according to the miles of mains multiplied by the number of SCs. The geographic and utility size distributions of the responses shown in Figures 2.8 and 2.9 are a fair representation of water utilities in North America.

DATA ANALYSIS

Data from survey responses were entered into a Microsoft Access database file to facilitate storage and analysis of the information. This relational database consisted of 9 tables, each containing information from a specific portion of the survey instrument. Data from the tables were exported to Microsoft Excel spreadsheets for analysis. Since the number of utilities reporting permeation incidents and successful uses was relatively small and the case history information was complex, the database file was not used to store specific information related to permeation incidents and successful uses.

Permeation incidents are occurrences for which there was a customer complaint about taste, odor, or illness or for which laboratory data exceeded U.S. EPA maximum contaminant levels (MCLs) that could be attributed to permeation of pipes or gaskets by hydrocarbons.

Successful uses are occurrences for which the utility became aware that pipes or gaskets were in contact with hydrocarbon contamination but for which there were no customer complaints or laboratory data exceeding U.S. EPA MCLs that could be attributed to permeation.

Mains

The 151 utilities responding to the survey reported 83,360 miles of mains, and 70% of these reported having at least some plastic mains. PVC and ductile iron (DI) pipe accounted for 18% and 16% of miles of mains reported, respectively (Figure 2.10). PE accounted for only 0.18% of miles of mains reported. Respondents considered only 0.54% of mains to be at risk of permeation.

Permeation Incidents

Reports of permeation of water mains were rare, approximately 1 report per 14,000 miles of mains. Of the 6 reported permeation incidents, 3 involved gasoline, 1 involved chlorinated solvents, and 2 involved unknowns. The pipe materials involved in permeation incidents were PVC (4), asbestos cement (AC) (1), and cast iron (CI) (1) (Figure 2.11).

Successful Uses

Reports of successful uses were infrequent, about 1 report per 1,800 miles of mains. Nearly all successful uses involved gasoline or chlorinated solvents. Isolated cases involving diesel fuel, polycyclic aromatic hydrocarbons, and acetone were also reported. The pipe materials involved in successful uses were DI (32), PVC (9), CI (5), and steel (1) (Figure 2.11).

Utility Size (miles of mains x No. of service connections): ● >10^8 (largest)
● 10^8–10^7 (large) ● 10^7–10^6 (medium) • 10^6–10^5 (small) • <10^5 (smallest)

Figure 2.8 Location distribution of the responding utilities

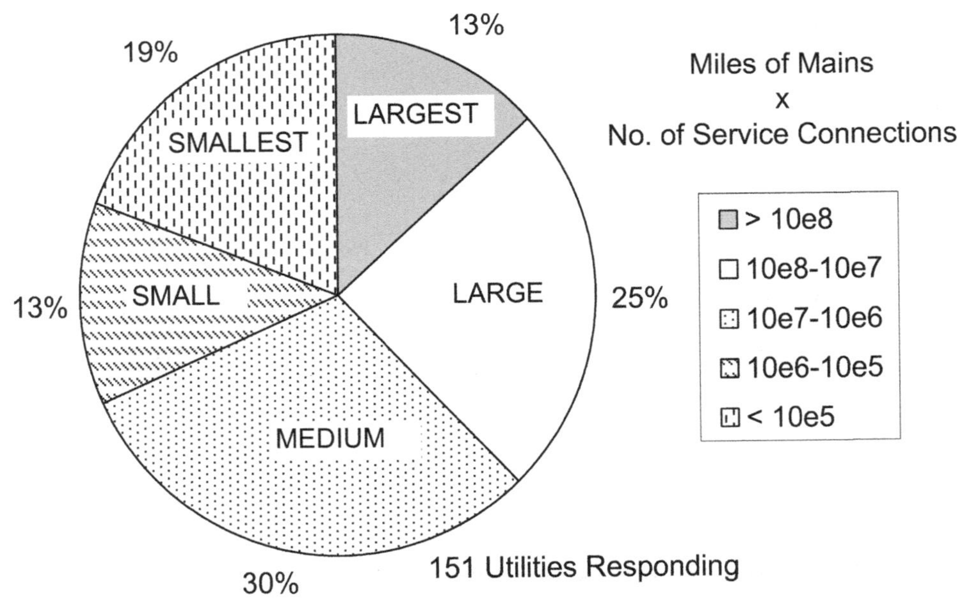

Figure 2.9 Relative size distribution of the responding utilities

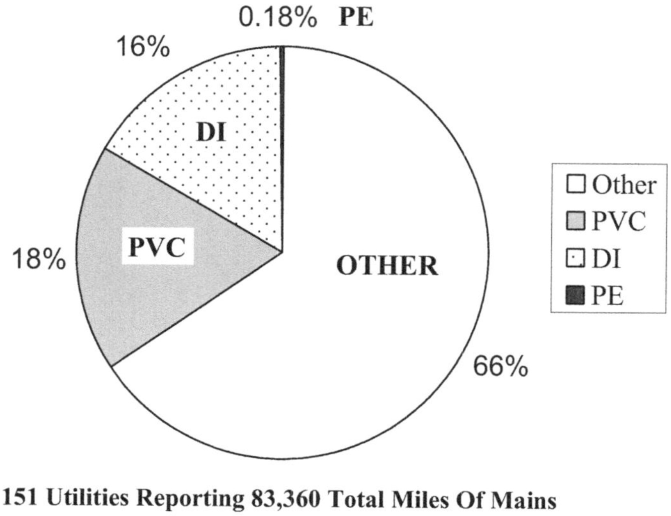

Figure 2.10 Relative distribution of pipe materials used in the reported mains (PVC – polyvinyl chloride, DI – ductile iron, PE – polyethylene, Other – cast iron (CI), steel, asbestos cement (AC) and concrete)

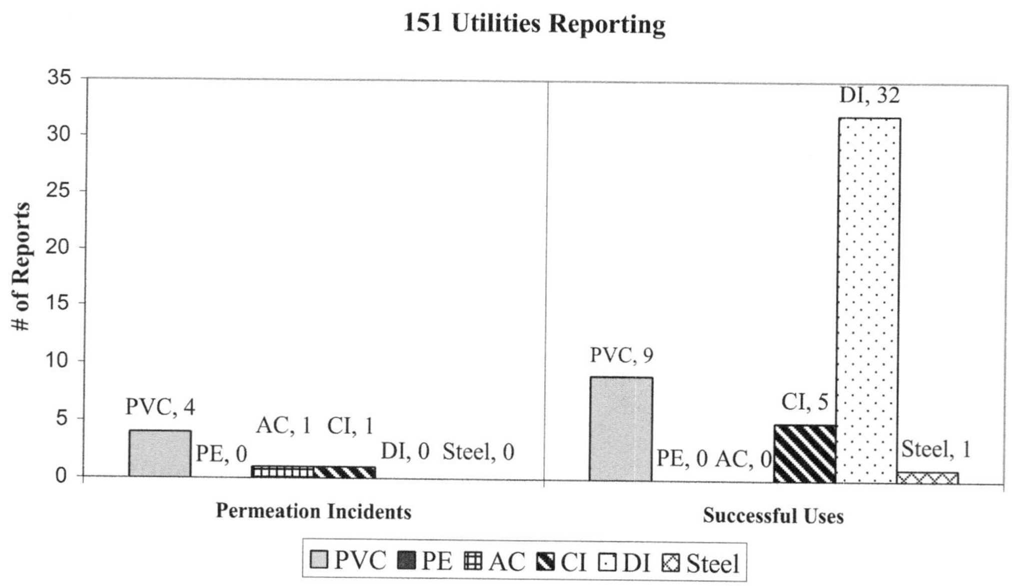

Figure 2.11 Permeation incidents and successful uses for the reported mains (Permeation incidents resulted in customer complaints or exceeded MCLs. Successful uses were occurrences of known contamination that did not result in incidents)

Protocols for Use of Pipes and Gaskets in Contaminated Soils

Protocols for use of mains in contaminated soils were reported by 19% of responding utilities (Figure 2.12). Use of DI with resistant gaskets (either NBR or FKM) was the most commonly reported protocol (10), followed by use of DI without specifying the gaskets (8), use of "no plastic" (4), and steel (2). Other protocols were reported by 4 utilities, included casing, "double wrapping", and "specifying alternate materials." Utilities specifying "no plastic" presumably use DI. While utilities reported specifying FKM gaskets in contaminated areas, manufacturers report that sales of FKM gaskets are extremely low relative to NBR gaskets, the most commonly used resistant gasket material (Corbett, 2006). No utility reporting a permeation incident reported having such a protocol.

Service Connections

The 151 utilities responding to the survey reported approximately 5,444,220 SCs and 49% of these reported allowing plastic SCs. PE and PVC pipe accounted for 6% and 5% of reported SCs, respectively (Figure 2.13). Respondents considered only 0.31% of SCs to be at risk of permeation. (Note: In the U.S., most service connections are not owned by the water utility but are the responsibility of the customers).

Permeation Incidents

Reports of permeation of PVC or PE SCs were about 0.9 reports per million SCs (Figure 2.14). Reports of permeation of all types of SCs were about 8 reports per million SCs. Of the 44 reported permeation incidents, 36 involved permeation of PB by gasoline. The permeation incidents involving PE (3) and PVC (2) SCs were with gasoline. There were 2 reports of permeation of DI/SBR SCs by chlorinated solvents. There was one report of permeation of an EPDM gasket on a copper SC by asphalt solvents which had entered the curb box. No reports of permeation of SCs due to use of termiticides were received. In all cases, the corrective action for permeation of plastic service lines was replacement with copper.

Successful Uses

Reports of successful uses of plastic SCs were about 0.4 reports per million total SCs (excluding the large number of reports involving copper SCs) (Figure 2.14). There were 2 reports of successful uses involving plastic pipe, one each involving PE and PB pipe. While DI accounted for only 0.11% of reported service connections, there were 13 reports of successful uses of DI SCs in contaminated soils. Similar to mains, utilities favor the use of DI pipe, often with FKM gaskets, for large services in hydrocarbon-contaminated areas.

Quantitative Data

Of the 25 utilities reporting incidents or successful uses, 10 had laboratory data relating to 18 cases. Table 2.1 shows the availability of laboratory data for permeation incidents and successful uses involving mains and services, and for the three sample matrices: soil, ground water, and potable water. There were more reported permeation incidents for SCs than for mains.

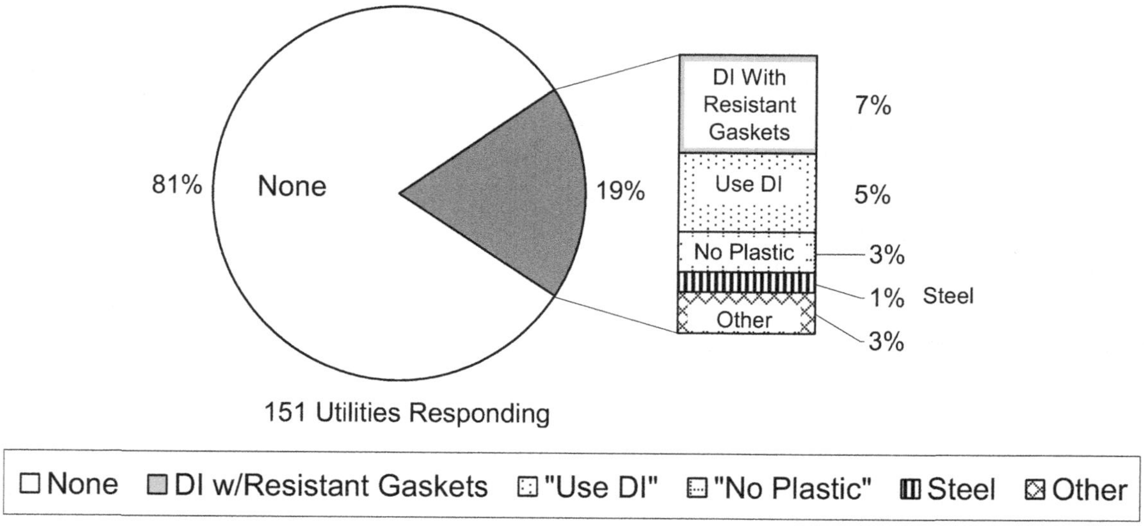

Figure 2.12 Protocols for use of mains in contaminated soils for responding utilities

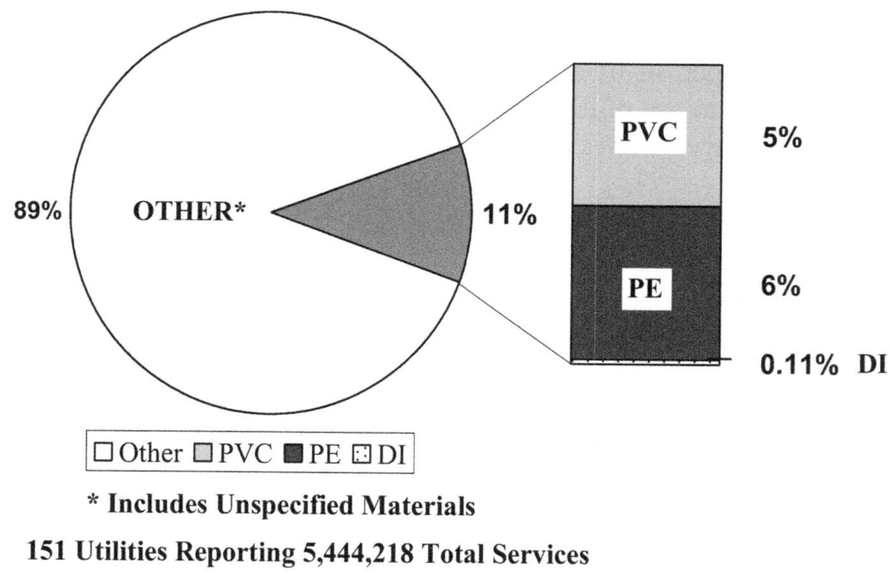

Figure 2.13 Relative distribution of pipe materials for the reported service connections

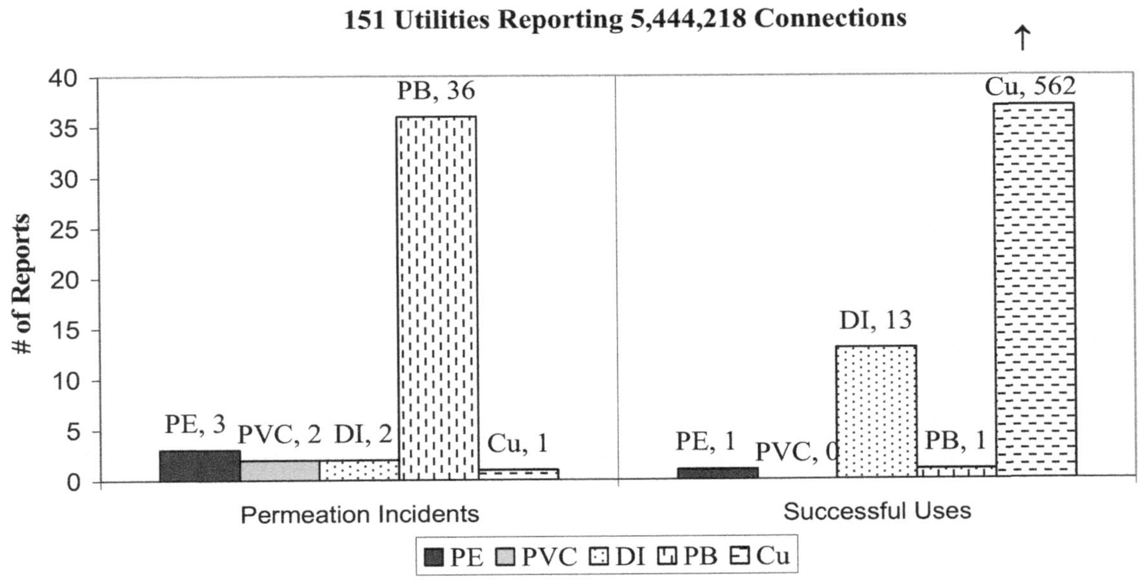

Figure 2.14 Permeation incidents and successful uses for the reported service connections (Permeation incidents resulted in customer complaints or exceeded MCLs. Successful uses were occurrences of known contamination that did not result in incidents.

Table 2.2 anonymously presents all analytical data reported by utilities, indicating the sample matrix and the pipe and gasket materials involved. Utilities are identified only by relative size and the U.S. EPA regions in which they are located. Case history numbers in the first column refers to the case histories listed in Appendix A.

There were no reports of soil or groundwater data associated with permeation incidents involving mains, but there were several reports of such data associated with successful uses of mains in contaminated soils. Contaminants reported were petroleum hydrocarbons (such as BTEX, polycyclic aromatic hydrocarbons (PAHs)) and chlorinated solvents (such as TCE, tetrachloroethylene (PCE), dichloroethylene (DCE) and vinyl chloride (VC)). Reported BTEX concentrations in groundwater ranged from as high as 51.7 mg/L to as low as 15 μg/L with concentrations in soils as high as 162 mg/kg. Concentrations of individual chlorinated solvent compounds in groundwater were typically less than 1 mg/L. However, there was a reported soil PCE concentration of 1,560 mg/kg. The only reports of successful uses of SCs in contaminated soils involved DI pipe, and the data are included in Table 2.3 with the data for mains.

Table 2.3 presents the upper range of contaminant concentrations that were reported in connection with the successful use of mains (and DI services) in contaminated soils. There is no presumption, however, that these are the upper limits of contamination in which these materials can be successfully used. Table 2.4 presents the range of contaminant concentrations that were reported in connection with permeation incidents involving SCs. There is no presumption, however, that these materials are safe from permeation at contaminant concentrations below those shown in Table 2.4. The data in Table 2.3 and 2.4 represent commonly encountered field conditions and/or realistic conditions of contamination in which PE/PVC pipes and pipe gaskets were successfully used, or which resulted in permeation incidents.

Gas Industry Experience with Permeation

Contacts with the gas industry (Lauder 2004, Ingels 2004) revealed that gaskets are no longer used in the gas industry and that PE gas pipes are customarily joined by fusion. While permeation of plastic gas pipe by petroleum-based hydrocarbons has occurred, it has not been known to cause any direct material failures or to have resulted in customer complaints or degradation of the delivered product. Permeation by hydrocarbons does cause a problem with making repairs or connections to sections of PE pipes that have been swollen by hydrocarbons. Connections to such pipes are unsuccessful because of non-uniformities and bubbles in the fusion area. The American Gas Association and the U.S. Department of Transportation jointly publish a database of pipe failures entitled "Plastic Pipe Data Collection." That database contains no reports of pipe failures due to permeation by petroleum-based hydrocarbons.

Table 2.1
Availability of laboratory data for permeation incidents and successful uses involving mains and services

Mains			Services		
Sample Matrix	Permeation Incidents	Successful Uses	Sample* Matrix	Permeation Incidents	Successful Uses
Soil	-	5	Soil	4	1
GW	-	6	GW	3	2
PW	1	6	PW	5	2

* GW = groundwater, PW = potable water

Table 2.2
Laboratory data reported by water utilities

Case History	Utility Size*/ EPA Region	Sample Matrix[+]	Analyte[ψ]	Concentration	Units	Successful Use (Y-Yes, N- No)	Main or Service	Materials
F	L/7	Soil	Tetrachloroethylene	0.036 - 3.60	mg/kg	Y	Main	DI/SBR
F	L/7	GW	Tetrachloroethylene	0.90 - 15.4	mg/L	Y	Main	DI/SBR
F	L/7	PW	Tetrachloroethylene	ND[ς] - 0.00675	mg/L	Y	Main	DI/SBR
F	L/7	Soil	Tetrachloroethylene	0.036 - 3.60	mg/kg	N	Service	DI/SBR
F	L/7	GW	Tetrachloroethylene	0.90 - 15.4	mg/L	N	Service	DI/SBR
F	L/7	PW	Tetrachloroethylene	ND - 0.00675	mg/L	N	Service	DI/SBR
J	L/7	Soil	2-methylnaphthalene	0.37	mg/L	Y	Main	CI/SBR
J	L/7	GW	Acetone	0.720 - 0.790	mg/L	Y	Main	CI/SBR
J	L/7	GW	total xylene	0.055 - 0.180	mg/L	Y	Main	CI/SBR
J	L/7	GW	Naphthalene	0.059	mg/L	Y	Main	CI/SBR
J	L/7	GW	2-methylnaphthalene	0.089	mg/L	Y	Main	CI/SBR
J	L/7	GW	Phenanthrene	0.045	mg/L	Y	Main	CI/SBR
J	L/7	GW	total extractable hyd.	24.8	mg/L	Y	Main	CI/SBR
J	L/7	GW	Diesel	10.1	mg/L	Y	Main	CI/SBR
J	L/7	GW	Gasoline	12.8	mg/L	Y	Main	CI/SBR
J	L/7	GW	motor oil	1.9	mg/L	Y	Main	CI/SBR
J	L/7	GW	Benzene	0.0258	mg/L	Y	Main	CI/SBR
J	L/7	GW	Toluene	0.0144	mg/L	Y	Main	CI/SBR
J	L/7	GW	Ethylbenzene	0.0771	mg/L	Y	Main	CI/SBR
J	L/7	PW	Toluene	0.0001-0.0002	mg/L	Y	Main	CI/SBR
J	L/7	PW	Ethylbenzene	0.0001	mg/L	Y	Main	CI/SBR
J	L/7	PW	m/p-xylene	0.0002	mg/L	Y	Main	CI/SBR
J	L/7	PW	o-xylene	0.0001	mg/L	Y	Main	CI/SBR
J	L/7	PW	Benzene	ND	mg/L	Y	Main	CI/SBR
J	L/7	Soil	2-methylnaphthalene	0.37	mg/kg	N	Service	CI/SBR
J	L/7	GW	2-methylnaphthalene	0.089	mg/L	N	Service	CI/SBR
J	L/7	Soil	2-methylnaphthalene	0.37	mg/kg	Y	Service	DI/FKM
J	L/7	GW	2-methylnaphthalene	0.089	mg/L	Y	Service	DI/FKM

(continued)

Table 2.2 (Cont'd.)
Laboratory data reported by water utilities

Case History	Utility Size/ EPA Region	Sample Matrix	Analyte	Concentration	Units	Successful Use (Y- yes, N – No)	Main or Service?	Materials
E	L/6	Soil	total BTEX	7.7	mg/kg	N	Service	PE
E	L/6	Soil	TPH	24	mg/kg	N	Service	PE
E	L/6	GW	total BTEX	57.7	mg/L	N	Service	PE
E	L/6	GW	TPH	45	mg/L	N	Service	PE
Y	M/7	Soil	Tetrachloroethylene	1,560	mg/kg	Y	Main	CI/Pb
Y	M/7	GW	Tetrachloroethylene	0.44	mg/L	Y	Main	CI/Pb
Y	M/7	PW	Tetrachloroethylene	ND	mg/L	Y	Main	CI/Pb
Y	M/7	Soil	Tetrachloroethylene	59.0	mg/kg	Y	Main	CI/Pb
Y	M/7	PW	Tetrachloroethylene	ND	mg/L	Y	Main	CI/Pb
G	M/7	GW	Total VOC (chlorinated)	0.715	mg/L	Y	Main	DI/FKM
G	M/7	PW	Total VOC (chlorinated)	ND	mg/L	Y	Main	DI/FKM
G	M/7	GW	Total VOC (chlorinated)	0.715	mg/L	Y	Service	DI/FKM
G	M/7	PW	Total VOC (chlorinated)	ND	mg/L	Y	Service	DI/FKM
Z	M/7	GW	Total VOC (chlorinated)	0.014	mg/L	Y	Main	DI/SBR
Z	M/7	PW	Total VOC (chlorinated)	ND	mg/L	Y	Main	DI/SBR
Z	M/7	GW	Total VOC (chlorinated)	0.014	mg/L	Y	Service	DI/SBR
Z	M/7	PW	Total VOC (chlorinated)	ND	mg/L	Y	Service	DI/SBR
D	XL/5	Soil	total BTEX	162	mg/kg	Y	Main	welded steel
V	XL/6	PW	Benzene	76	µg/L	N	Service	PE
V	XL/6	PW	Toluene	59	µg/L	N	Service	PE
V	XL/6	PW	2-chlorotoluene	266	µg/L	N	Service	PE
V	XL/6	PW	4-chlorotoluene	131	µg/L	N	Service	PE
V	XL/6	PW	Chlorobenzene	0.4	µg/L	N	Service	PE
V	XL/6	PW	1,2-dichlorobenzene	12.5	µg/L	N	Service	PE
V	XL/6	PW	1,2,4-trichlorobenzene	3	µg/L	N	Service	PE
V	XL/6	PW	1,2,3-trichlorobenzene	0.4	µg/L	N	Service	PE
V	XL/6	PW	Ethylbenzene	29.4	µg/L	N	Service	PE
V	XL/6	PW	o-xylene	0.4	µg/L	N	Service	PE

(continued)

Table 2.2 (Cont'd.)
Laboratory data reported by water utilities

Case History	Utility Size/ EPA Region	Sample Matrix	Analyte	Concentration	Units	Successful Use (Y- yes, N – no)	Main or Service	Materials
V	XL/6	PW	isopropyl benzene	2.1	μg/L	N	Service	PE
V	XL/6	PW	t-butylbenzene	0.3	μg/L	N	Service	PE
V	XL/6	PW	s-butylbenzene	0.4	μg/L	N	Service	PE
V	XL/6	PW	n-butylbenzene	15.6	μg/L	N	Service	PE
U	XL/10	PW	total BTEX	0.015	mg/L	N	Service	PB
T	XL/4	Soil	n-butylbenzene	0.34	mg/kg	N	Service	Cu/EPDM
K	L/9	Soil	Acetone	44,700	mg/kg	N	Service	PB
K	L/9	PW	Acetone	16	mg/L	N	Service	PB
AA	L/9	PW	TPH	0.05	mg/L	Y	Main	PVC/SBR
T	XS/8	GW	Tetrachloroethylene	250	μg/L	Y	Main	PVC/FKM
T	XS/8	GW	trichloroethylene	73	μg/L	Y	Main	PVC/FKM
T	XS/8	GW	cis-dichloroethylene	241	μg/L	Y	Main	PVC/FKM
T	XS/8	GW	vinyl chloride	17	μg/L	Y	Main	PVC/FKM
T	XS/8	GW	Tetrachloroethylene	262	μg/L	Y	Main	PVC/FKM
T	XS/8	GW	trichloroethylene	119	μg/L	Y	Main	PVC/FKM
T	XS/8	GW	cis-dichloroethylene	266	μg/L	Y	Main	PVC/FKM
T	XS/8	GW	vinyl chloride	13	μg/L	Y	Main	PVC/FKM
T	XS/8	GW	Tetrachloroethylene	96	μg/L	Y	Main	PVC/FKM
T	XS/8	GW	trichloroethylene	13	μg/L	Y	Main	PVC/FKM
T	XS/8	GW	vinyl chloride	12	μg/L	Y	Main	PVC/FKM
T	XS/8	GW	Tetrachloroethylene	952	μg/L	Y	Main	PVC/FKM
T	XS/8	GW	trichloroethylene	133	μg/L	Y	Main	PVC/FKM
T	XS/8	GW	cis-dichloroethylene	386	μg/L	Y	Main	PVC/FKM
T	XS/8	GW	vinyl chloride	34	μg/L	Y	Main	PVC/FKM
H	M/9	PW	1,2-dichlorobenzene	5 - 8	μg/L	N	Main	2-in. PVC$^{\phi}$
H	M/9	PW	1,4-dichlorobenzene	2	μg/L	N	Main	2-in. PVC$^{\phi}$

* Utility size (XL – largest, L – large, M – medium, S – small, XS – smallest)
$^{+}$ GW – groundwater, PW – potable water
$^{\varsigma}$ ND – non detect
$^{\Psi}$ TPH – Total Petroleum Hydrocarbons, VOC – Volatile organic compounds
$^{\phi}$ jointing unknown

Table 2.3
Upper range of contaminant concentrations associated with successful use of mains in contaminated soils

Materials*	Soil	Groundwater
SBR Gaskets w/DI or CI Pipe	3.6 mg/kg tetrachloroethylene	15.4 mg/L tetrachloroethylene 0.30 mg/L Total BTEX 25.8 μg/L benzene
DI/SBR Service Line	-	25.8 μg/L benzene 14 μg/L VOC**
PVC/FKM	-	0.952 mg/L tetrachloroethylene 0.133 mg/L trichloroethylene 0.386 mg/L cis-dichloroethylene 0.034 Vinyl chloride 1.505 mg/L total chlorinated***
DI/FKM Service Lines	-	0.30 mg/L Total BTEX 25.8 μg/L benzene 0.30 mg/L Total BTEX 25.8 μg/L benzene 0.715 mg/L VOC**

* Uses are for mains, except where noted
** VOC – volatile organic compounds - contaminants were unspecified chlorinated solvents
*** Sum of PCE, TCE, cis-DCE, & VC

Table 2.4
Range of contaminant concentrations associated with permeation incidents involving service connections

Materials*	Soil	Groundwater
SBR Gaskets w/DI or CI Pipe	0.036 - 3.6 mg/kg tetrachloroethylene	0.9 - 15.4 mg/L tetrachloroethylene 0.17 - 0.30 mg/L Total BTEX 25.8 μg/L benzene
PE Pipe	7.7 mg/kg Total BTEX 24 mg/kg TPH	57.7 mg/L Total BTEX 45 mg/L TPH

Case Histories

The case histories presented in Appendix A were developed from the information in the survey responses and from conversations with water utility officials. Case histories are presented here in an anonymous form, with all names, dates, and geographical information omitted. Laboratory data has been included where available. Statements in the case histories preceded by "presumably" are inferences made by the authors, not utility officials.

CONCLUSIONS

Eighteen percent of the total miles of mains reported were PVC, 16% were ductile iron, 66% reported others which included cast iron, steel, asbestos cement and concrete pipes. Only 0.18% of the total mains reported were PE pipes. With respect to service lines, 49% of responding utilities reported allowing use of plastic SCs. PE and PVC pipe accounted for 6% and 5% of reported SCs, respectively.

The overall impact of permeation on the water industry appears to be relatively small. Reports of permeation incidents involving potable water distribution are rare. Nevertheless, every permeation incident that does occur is a potential threat to public health and customer satisfaction.

Although 19% of responding utilities reported that they had a protocol for use of pipe and gasket materials in contaminated areas, none of the utilities reporting permeation incidents had such a protocol. Utilities reporting protocols mainly specify DI pipe, usually with resistant gaskets (especially FKM), in contaminated areas; none specified plastic pipes.

The typical response of a water utility to a permeation incident is to fix the problem as quickly as possible to minimize public health risk and customer inconvenience. The source of the permeation is often obvious and samples that reflect the "as found" condition of the soil, ground water, and potable water are seldom collected. The data suggest that service lines are more vulnerable to permeation incidents than mains, presumably because of stagnation periods. The quantitative data associated with permeation incidents and successful uses reported in this survey was not sufficient to allow conclusions to be drawn regarding the thresholds of contamination that would require a utility to choose more resistant pipe or gasket materials. Such decisions must necessarily be based on the laboratory data presented elsewhere in this report.

CHAPTER 3
POLYETHYLENE PIPES - LABORATORY STUDIES UNDER SIMULATED ENVIRONMENTAL CONDITIONS

OVERVIEW

Polyethylene (PE) pipes are widely used in water service connections, mainly due to their excellent abrasion, impact and corrosion resistance. However, there is strong evidence in the literature and from field studies to suggest that organic contaminants permeate PE pipes and adversely affect the quality of drinking water in water distribution systems. According to a survey completed in the 1980s (Thompson and Jenkins 1987), PE pipes were involved in 39 % of total incidents of drinking water contamination resulting from permeation. Contamination by aromatics and chlorinated solvents in drinking water has been reported as a result of permeation through PE pipes after a period of water stagnation in the pipe (Holsen et al. 1991a).

Relatively high permeability of organic compounds through PE pipe is attributed to the structural characteristics of PE. PE is characterized as a semi-crystalline polymer, having both crystalline and amorphous regions. The crystalline zones act as impermeable barriers for sorption and diffusion, while the non-crystalline matrix is readily permeable since the polymeric chains in the amorphous areas are relatively "mobile" (Naylor 1989).

Although it is well known that organic compounds permeate PE pipes readily, significantly less effort has been devoted to simulate field contamination conditions and thus there are few data which can be applied to predict the permeation behavior in the real world. Field contamination conditions are generally characterized by complex mixtures of organic chemicals distributed among aqueous, solid, and gas phases and non-aqueous phase liquids (NAPL), which implicitly determine the effective availability of these compounds for permeation through pipe materials. For example, gasoline released into the environment may be trapped by capillary force in the soil of the vadose zone as NAPL (Zhou and Crawford 1995), volatilize and form hydrocarbon vapors in the vadose zone (Jutras et al. 1997), or may migrate downward to the water table, resulting in the pooling of gasoline and dissolution of hydrocarbons into aqueous phase (Cline et al. 1991). Therefore, permeation can occur either from the vapor, aqueous, or NAPL phases and plastic pipes buried in both the vadose and saturated zones are susceptible to permeation.

The aim of this study was to investigate the permeation of petroleum-based aromatic compounds (benzene, toluene, ethylbenzene, m-xylene, and o+p-xylene (BTEX)) through PE pipes under simulated field conditions including subsurface gasoline spills, gasoline-contaminated groundwater, and gasoline-contaminated soil with varied levels of contamination. For each specific testing condition, the permeation fluxes of BTEX compounds through PE pipe were quantified and the diffusion coefficients of BTEX compounds in PE pipe were estimated. Correlations between the steady state permeation rates and the bulk concentrations were established and empirical equations developed. The results served as a basis for predicting the permeation behavior of BTEX through PE pipe at contaminant levels typically encountered in the environment.

MATERIALS AND METHODS

Materials

One-inch PE pipe (high density, SIDR 9 IPS NSF PW PE3408 ASTM-D2239-81) was obtained from a PE pipe manufacturer. Premium gasoline was purchased from a local store and the concentrations of benzene, toluene, ethylbenzene, m-xylene, and o+p-xylene in the gasoline were determined by gas chromatography to be 19.8 g/L, 75.9 g/L, 14.7 g/L, 33.7 g/L, and 32.5 g/L, respectively. BTEX compounds were roughly 25% by volume of the gasoline, which was slightly higher than 19 - 24% for commercial gasoline reported by Cline et al. (1991). Three types of soils, silica sand, organic topsoil, and a sand-topsoil mixture were used in the experiments. Silica sand (Granusil 4030) was purchased from UNIMIN Corporation, Portage WI. Organic topsoil, purchased from a local store, was air dried and sieved to pass a 2 mm sieve. Another soil was prepared by mixing approximately equal parts of silica sand and dried, sieved, organic topsoil with a Hobart mixer. Triplicate samples were taken randomly from each prepared soil and analyzed for their important physical and chemical properties according to the Methods of Soil Analysis (Klute, 1986; Page, 1986). The results are summarized in Table 3.1.

Experimental Apparatus

Experiments were conducted in a mini pipe-bottle apparatus (Figure 3.1), consisting of a 1-L glass bottle with the PE pipe mounted horizontally through holes drilled in the glass. The connections between pipe and bottle were sealed with Loctite® plastic epoxy and further covered by Loctite® epoxy putty (Henkel Technologies, Rocky Hill, CT). The ends of the pipes were sealed with Teflon® plugs. The Teflon® plug had a small hole that was plugged with a brass plug to allow filling and draining of the water inside the pipe with a glass syringe. The bottle was capped with a Teflon® -lined cap. The total surface area of PE pipe exposed to the bulk solution was approximately 81.9 cm^2. To simplify the experimental system and facilitate ease in handling the mini pipe-bottle apparatus, all experiments were conducted at room temperature (23 ± 1.5°C).

Experimental Procedures

Exposure of PE Pipes to Gasoline

One experiment investigated the permeation of BTEX through PE pipe exposed to pure premium gasoline. In this test, the pipe segment was first filled with deionized water and the bottle filled with silica sand and premium gasoline until the bottle was nearly full and a visible liquid level of gasoline appeared above the surface of the sand. The level of liquid gasoline in the bottle was monitored visually and there was no significant drop in the level during the experimental period of four weeks. Pipe-water samples were taken at regular intervals and analyzed for the presence of BTEX. The pipe was flushed with deionized water three times immediately after each sampling before fresh deionized water was added and the pipes sealed.

Table 3.1
Physiochemical characteristics of soils used for testing

Soil	Silica sand	Organic topsoil	Sand-topsoil mixture
Soil pH	6.6	6.2	6.3
Organic carbon content (%)	Below detection limit	5.1	1.9
Specific surface area (m^2/g)	0.6	115	36
Particle size distribution			
Sand (%)	99.7	93.9	97.2
Silt (%)	0.3	5.1	2.5
Clay (%)	0	1.0	0.4

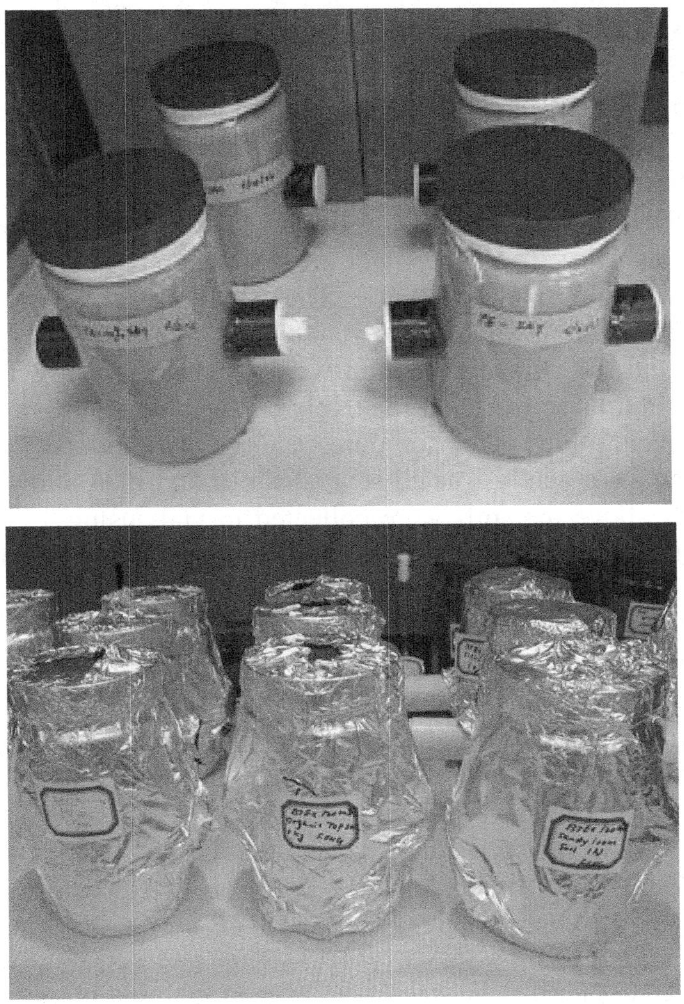

Figure 3.1 Pipe-bottle apparatus

Exposure of PE Pipes to Gasoline-Contaminated Groundwater

Aqueous gasoline-saturated solution was prepared by mixing 350 mL premium gasoline and 3.5 L deionized water in a 4-L glass bottle for 48 hours using a magnetic stirrer. Several preliminary experiments indicated that a mixing time of 48 hours was sufficient to produce an aqueous gasoline-saturated solution. A freshly saturated stock solution was prepared once every week throughout this study. The concentration of the dissolved BTEX was around 150 mg/L, which was slightly higher than the average value of 135 mg/L reported by Cline et al. (1991) for water saturated with gasoline of different grades. Benzene and toluene were the major aromatic compounds in the aqueous phase, accounting for nearly 90% of total BTEX. Three diluted aqueous solutions of varying concentrations, 50%, 10%, and 1% aqueous solutions of gasoline, were obtained by diluting the stock solution with deionized water at a ratio of 1:1, 1:9, 1:99, respectively. The range of BTEX concentrations in the four aqueous solutions (≈150 mg/L, ≈75 mg/L, ≈15 mg/L and ≈1.5 mg/L for 100%, 50%, 10% and 1% aqueous solutions of gasoline, respectively) typically reflect the concentration range of BTEX found in groundwater associated with gasoline release.

The mini pipe-bottle apparatuses were filled with silica sand until the bottle was nearly full. Sodium azide (1.5 g) was added and thoroughly mixed with the sand by rotating the bottle in order to inhibit potential biodegradation of BTEX. The bottle was loosely capped with a Teflon®-lined cap through which a small hole was drilled and a Teflon® tube installed, reaching to the bottom of bottle. Aqueous gasoline solution was then introduced into the sand through the Teflon® tubing, from bottom to top, using a Masterflex pump, until the bottle was full. The Teflon® tube was removed along with the cap and the bottle was then tightly capped with a Teflon®-lined cap with no hole. The final water level was approximately 2 cm (¾ inch) above the surface of the soil, with minimum headspace. The apparatuses were wrapped with aluminum foil to minimize photodegradation.

The silica sand and the aqueous solution were replaced each week with new sand and fresh aqueous gasoline solution to maintain relatively constant aqueous concentrations of BTEX in the bottle. BTEX concentrations in the soil pore water were periodically measured. To collect samples, the pipe-bottles were gently rotated several times and then allowed to sit undisturbed for 10 minutes. Liquids above the soils were collected in vials using a glass syringe and then centrifuged at 5,000 rpm for 10 minutes. These supernatant liquids were analyzed in the same manner as the pipe-water samples.

Exposure of PE Pipes to Unsaturated Gasoline-Contaminated Soil

Four levels of gasoline-contaminated soil were prepared by spiking 1 kg sand-topsoil mixture with known amounts of premium gasoline and then rotating the soil container for one week. The volumetric moisture content of the soil was adjusted to approximately 10% - which was above the water content of the soils equilibrated at a relative humidity > 98 % (Holsen et al. 1991b). The amounts of BTEX sorbed onto the four soils were determined using methanol extraction followed by gas chromatography. The total BTEX in the four soils were found to be 32, 89, 388 and 1,216 mg/kg dry soil, respectively. Toluene and xylene were the major aromatic compounds in the soil accounting for approximately 90% of the total BTEX.

In these experiments, gasoline-contaminated soils were packed into the pipe-bottle apparatus and were not mixed. Under those conditions, mass transfer limitations in the soil might have occurred when organic compounds were absorbed rapidly by the pipe wall, decreasing the concentration of organic compounds in the soil immediately adjacent to the pipe wall. Due to volatilization and diffusion, the concentration of BTEX surrounding the pipes may not have remained constant over the experimental period. During the experiments, the water inside the pipes was periodically drained for BTEX analysis and replenished with fresh deionized water. At the end of the experiments, soil samples were collected from the top layer (around 4 cm (1.5 ins.) above the pipe), middle layer (surrounding the pipe) and bottom layer (around 4 cm (1.5 ins.) below the pipe), and the residual BTEX in the soil was determined using methanol extraction followed by gas chromatography.

Impact of Soil Organic Matter on Permeation (Saturated Soil Experiments)

Pipes were buried in the three types of soils, silica sand, organic topsoil, and a sand-topsoil mixture, which were initially soaked with aqueous gasoline-saturated solutions. Soil pore water and pipe-water were collected over the experimental period and the BTEX concentrations measured. The experimental procedures were the same as described above in the section entitled "Exposure of PE pipes to Gasoline-Contaminated Groundwater", except that the soil and aqueous solution were not replenished. To compensate for the liquid loss (around 4 mL) from sampling and to maintain zero headspace, additional fresh gasoline-saturated water was rapidly added to each bottle following sampling. The additions did not affect the concentration of BTEX in soil pore water since the total capacity of soil pore water (around 400 mL) was greater than the volume of fresh gasoline-saturated water added.

Gas Chromatographic Determination of BTEX

BTEX in the samples was determined using a gas chromatograph (Tracor 540) equipped with a packed column (1.8 m (6 ft) × 2 mm; 1% SP1000 on 60/80 mesh Carbopack B), a photoionization detector, and an automated purge & trap concentrator (Tekmar LSC2/ALS). The method detection limits for benzene, toluene, ethylbenzene, m-xylene and o+p-xylene were 0.24 μg/L, 0.24 μg/L, 0.26 μg/L, 0.29 μg/L, and 0.53 μg/L, respectively.

Data Processing

The total mass of contaminant permeating PE pipe (mass/surface area) between sampling periods was estimated based on the contaminant concentration in the pipe water, the pipe volume and the pipe surface area. Using a plot of cumulative mass permeated per unit area, Q (μg/cm^2), versus time, the intersect with the time (t) axis of the straight line drawn through the $Q(t)$ data at steady state conditions, is defined as the time lag, T_L. The diffusion coefficients (D) for the pipe-contaminant systems can be estimated using the slope of the permeation curve or by using the time lag method.

The time lag method is an estimation technique based on equation 3.1, below which is derived from Fick's diffusion equation. The total quantity of chemical diffusing into a hollow cylinder ($Q(t)$) as a function of time (t), is based on the following simplifying assumptions: (i) the outer concentration of contaminant remains constant (C_0); (ii) the initial concentration of the contaminant in the polymer is zero; and (iii) the inner concentration of the contaminant is kept at zero.

$$\frac{Q(t)}{\pi C_o} = \frac{2(Dt - T_T)}{ln(a/b)} - 4\sum_{n=1}^{\infty} \frac{J_o(a\alpha_n)J_o(b\alpha_n)exp(-D\alpha_n^2 t)}{\alpha_n^2\{J_o^2(a\alpha_n) - J_o^2(b\alpha_n)\}} \tag{3.1}$$

where a = inner radius (cm)

b = outer radius (cm)

D = diffusion coefficient (cm^2/s)

$J_0(x)$ = Bessel function of the first kind of order zero

a_n = Positive roots of

$$U_0(a\alpha_n) = J_0(a\alpha_n)Y_0(b\alpha_n) - J_0(b\alpha_n)Y_0(a\alpha_n) = 0 \tag{3.2}$$

where $Y_0(x)$ = Bessel function of the second kind of order zero

T_L, the intercept made on the time axis by the asymptote to the plot of the quantity diffused with time, and by setting $Q_t = 0$, equation 3.2 can be simplified to the following equation (Crank, 1975):

$$T_L = \frac{b^2 - a^2 + (a^2 + b^2)\ln(a/b)}{4D\ln(a/b)} \tag{3.3}$$

As suggested by Barrer (1951), this intercept (referred to as the "time lag") provides a means of estimating the diffusion coefficient D of a chemical through a material in the form of a hollow cylinder.

RESULTS AND DISCUSSION

Exposure of PE Pipes to Gasoline

The cumulative masses permeated per unit area for BTEX compounds are plotted in Figure 3.2. The permeation of BTEX through PE pipe exposed to gasoline was extremely rapid. Of the five compounds of interest, benzene was the first compound detected with a breakthrough time of approximately one week. When breakthrough was detected, the concentration of benzene measured in the pipe-water exceeded the maximum contaminant level (MCL) of 5 µg/L. Toluene breakthrough lagged just slightly behind that of benzene, while ethylbenzene and xylene were detected later (approximately 15 days) and at much lower concentrations. Benzene and toluene accounted for nearly the entire total BTEX permeated into the pipe-water. This is mainly due to the higher solubility of benzene in water and the higher mass fraction of toluene in gasoline. Similar permeation behavior of BTEX (i.e., the order of breakthrough time and the relative permeation concentrations) was found in gasketed ductile iron (DI) pipe joints exposed to gasoline (Glaza and Park 1992; Chapter 6 of this report).

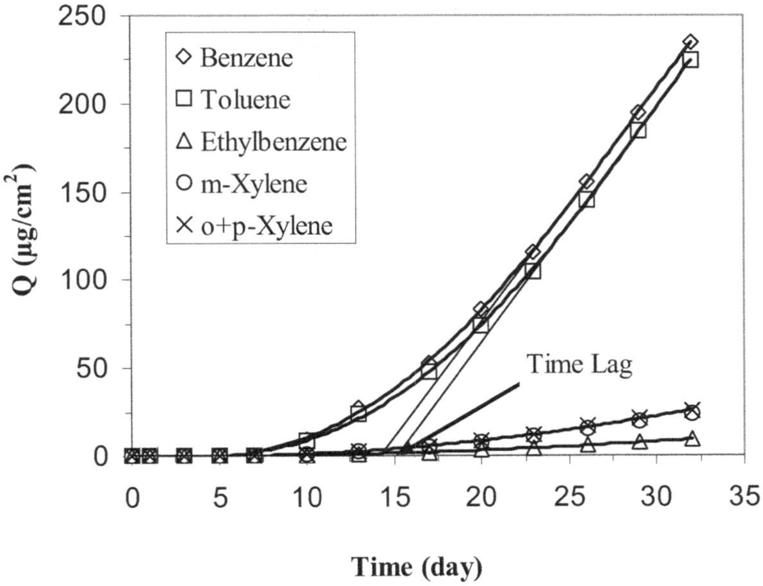

Figure 3.2 Cumulative mass permeated per unit area for BTEX compounds in PE pipe exposed to premium gasoline (Benzene, 19.8 g/L; toluene, 75.9 g/L; ethylbenzene, 14.7 g/L; m-xylene, 33.7 g/L; o+p xylene, 32.5 g/L)

The T_L for benzene and toluene were estimated to be 14.1 days and 15.2 days, respectively. Based on the time lag method (equation 3.3), the corresponding diffusion coefficients were estimated to be 1.3×10^{-8} cm^2/s and 1.1×10^{-8} cm^2/s, respectively. Under steady state permeation conditions, the concentrations of BTEX in the pipe water at the end of the 3-day sampling intervals were found to be very close to the BTEX concentrations in aqueous gasoline-saturated solution. Apparently, the pipe material was saturated with gasoline, resulting in BTEX concentrations in the pipe-water that were close to the saturation limit of gasoline in water.

Exposure of PE Pipes to Gasoline-Contaminated Groundwater

The cumulative mass permeated per unit area for BTEX compounds at four levels of contamination are plotted in Figure 3.3. The measured concentrations of BTEX in soil pore water at four levels of contamination were slightly lower than the expected concentrations (100%, 50%, 10% and 1%), mainly due to sorption by the pipe and volatilization loss.

As shown in Figure 3.3, benzene and toluene rapidly permeated through PE pipe, even from the most dilute solution. As in the pure gasoline exposure experiments, benzene was the first compound to be detected in the pipe-water, probably due to the high concentration of benzene in the simulated ground water and its smaller molecular diameter. Once breakthrough occurred, the level of benzene in the pipe-water quickly exceeded its MCL. The breakthrough of toluene was slightly behind that of benzene, while ethylbenzene and xylene were detected later (approx. 41 days for both compounds for gasoline-saturated water) and at much lower concentrations (breakthrough data not shown). Figure 3.3 also demonstrated that the

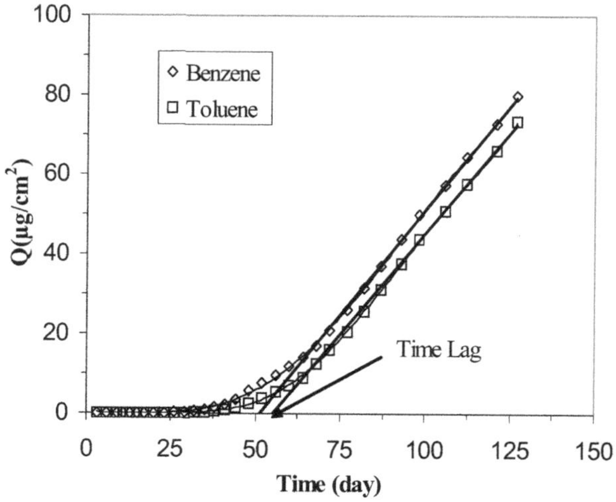

≈ 100% Saturated: Total BTEX, 136.6±13.0 mg/L; benzene, 67.5±4.9 mg/L; toluene, 56.2 ± 4.9 mg/L

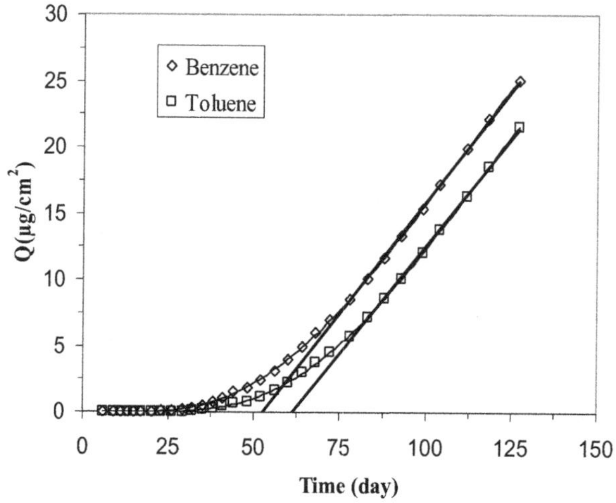

≈ 50% Saturated: Total BTEX, 63.9±5.7 mg/L; benzene, 31.2±2.9 mg/L; toluene, 26.3±2.4 mg/L

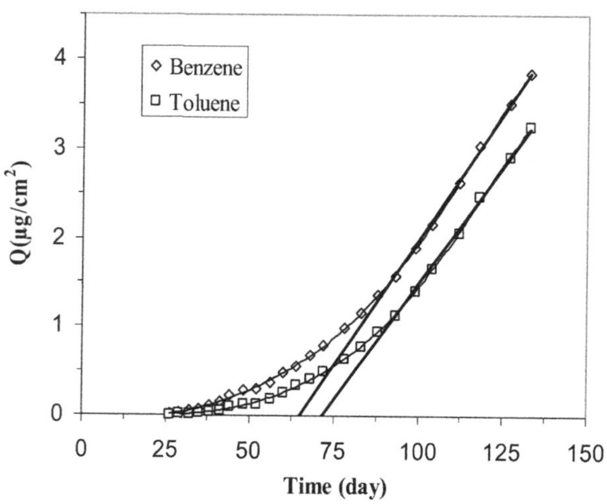

≈ 10% Saturated: Total BTEX, 12.6±1.1 mg/L; benzene, 6.0±0.6 mg/L; toluene, 5.2±0.4 mg/L

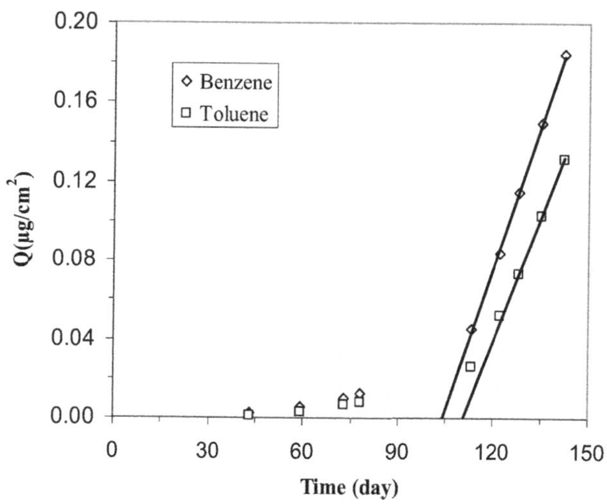

≈ 1% Saturated: Total BTEX, 1.2±0.6 mg/L; benzene, 0.6±0.3 mg/L; toluene, 0.5±0.3 mg/L

Figure 3.3 Cumulative mass permeated per unit area for benzene and toluene in PE pipes exposed to gasoline-contaminated water solutions

breakthrough time and the lag time increased with lower BTEX groundwater contamination, which implied that higher bulk groundwater hydrocarbon concentrations would lead to earlier breakthrough.

Diffusion coefficients (D) of benzene and toluene at four levels of contamination were estimated using the time lag method (equation 3.3). Diffusion coefficients for benzene and toluene ranged from 1.8×10^{-9} to 3.6×10^{-9} and 1.8×10^{-9} to 3.3×10^{-9} cm^2/s, respectively and are summarized in Table 3.2 for the different bulk exposure concentrations. Included in Table 3.2 for comparison purposes are diffusion coefficients from other studies.

As shown in Table 3.2, diffusion coefficients from this study and other studies were of the same order of 10^{-9} cm^2/s, despite the different polymer composition and crystallinity, different concentration of contaminants, and different experimental conditions used. Diffusion coefficients of toluene obtained from this study using PE pipes (high density) were similar to those of Vonk's study (1985) using low density polyethylene (LDPE) pipes. Table 3.2 also indicates that diffusion coefficients were concentration-dependent and increased with an increase in bulk exposure concentration. This concentration effect may be attributed to increased mobility of polymer segments resulting from increases in the average free volume in the polymer caused by the presence of the diffusing contaminant. For each contamination level (approximately gasoline-saturated (100%), 50%, 10%, or 1% aqueous solutions of gasoline), insignificant differences between the diffusion coefficients of benzene and toluene were found, which agreed with the results of Sangam and Rowe (2001) but conflicted with those of Joo et al. (2004). In Joo's study, diffusion coefficients of benzene were found to be 1.5 times higher than that of toluene for a bulk concentration of 30 mg/L. One possible explanation for the nearly identical permeation of benzene and toluene detected in this study might be due to the synergistic effect of permeation. Most studies reported in Table 3.2 were based the permeation of a single organic compound in aqueous solution through the polymeric material while this study used gasoline-contaminated water which is a mixture of organic compounds and, therefore, the presence of benzene with toluene may enhance the diffusion of toluene through the HDPE pipes. Synergistic effects were also observed in the study by Park et al. (1991) where toluene permeation through PB pipe was considerably enhanced by the presence of TCE.

Based on the $Q(t)$ curve at steady state, the permeation rates, P_m, for the corresponding contamination levels (approximately 100%, 50%, 10% and 1% saturated aqueous solutions) were estimated to be 1.044, 0.344, 0.057, and 0.0046 μg/(cm2·day) for benzene and 1.011, 0.325, 0.052, and 0.0046 μg/cm2/day) for toluene. The estimated values showed that the permeation rates were strongly dependent on the bulk concentrations of contaminants in the soil pore water. The steady state permeation rates versus the bulk (external) concentrations were plotted as shown in Figure 3.4. Two empirical correlation equations were obtained for the steady state permeation rate, P_m [$\mu g/cm^2/day$)] and the bulk concentration, C_{bulk} in soil pore water (mg/L):

$$\begin{aligned} P_m &= 0.0079 C_{bulk}^{1.1323} \quad \text{for benzene} \\ P_m &= 0.0087 C_{bulk}^{1.1444} \quad \text{for toluene} \end{aligned} \tag{3.4}$$

Table 3.2
Comparison of diffusion coefficients obtained from this study with previous studies

Organic chemical	Materials	Exposure concentration (mg/L)	Diffusion coefficient ($\times 10^{-9}$ cm^2/s)	References
Toluene	HDPE pipe Wall thickness: 3.1 mm	56.2±4.9 26.3±2.4 5.2±0.4 0.5±0.3	3.4 3.0 2.6 1.8	This study
Toluene	LDPE pipe Wall thickness: 3.5 mm	91±16 49±11 31±5 2.9±1.3	4.7* 3.9* 3.9* 3.5*	Vonk (1985)
Toluene	HDPE geomembrane Thickness: 2.0 mm	30	3.3	Joo et al. (2004)
Toluene	HDPE geomembrane Thickness: 0.76 mm	100 50 10	5.3 3.6 3.2	Park et al. (1996)
Toluene	HDPE geomembrane Thickness: 2.0 mm	2	3.0	Sangam and Rowe (2001)
Benzene	HDPE pipe Wall thickness: 3.1 mm	67.5±4.9 31.2±2.9 6.0±0.6 0.6±0.3	3.6 3.5 2.9 1.8	This study
Benzene	HDPE geomembrane Thickness: 2.0 mm	30	8.5	Joo et al. (2004)
Benzene	HDPE geomembrane Thickness: 2.0 mm	2	3.5	Sangam and Rowe (2001)

* Calculated on the basis of the permeation data in the reference.

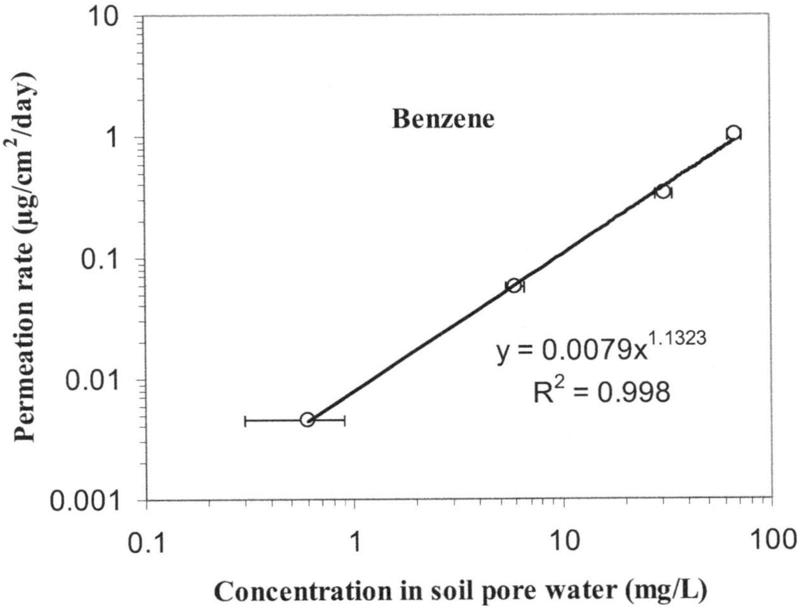

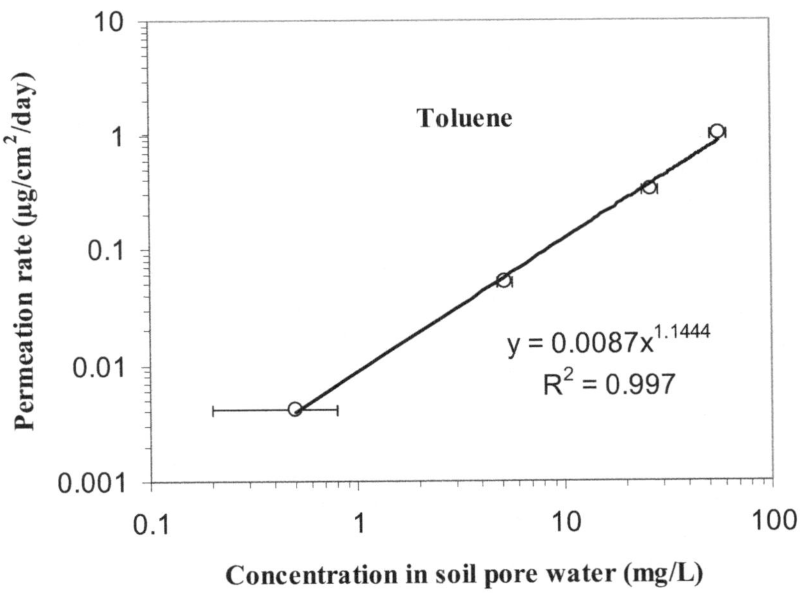

Figure 3.4 Correlation between the steady state permeation rates and the external bulk concentrations in soil pore water for benzene and toluene

The equations above can be used to predict the concentration of benzene or toluene in the pipe-water at a given bulk contaminant concentration after a period of stagnation. Stagnation, representing the worst-case scenario is typically found during the night in service lines to residences or office buildings. The concentration of contaminant in pipe-water (C_{pw}) can be estimated by:

$$C_{pw} = \frac{M}{V} = \frac{P_m \times \pi \times O.D \times L_C \times t}{\frac{1}{4} \times \pi \times I.D.^2 \times L_T} = \frac{4 \times f(C_{bulk}) \times O.D. \times L_C \times t}{I.D.^2 \times L_T} \quad (3.5)$$

where M = total mass of permeated contaminant (μg)

V = volume of water in the pipe (L)

P_m = steady permeation rate of the contaminant [μg/cm^2/day)], and is a function of the bulk concentration of contaminant in the soil pore water, $f(C_{bulk})$ (equation 3.4)

C_{bulk} = bulk concentration of contaminant in the soil pore water, C_{bulk} (μg/L)

t = period of stagnation (days)

L_C = length of contaminated pipe (cm)

L_T = total length of pipe (cm)

$I.D$ = inside diameter of pipe (cm)

$O.D.$ = outside diameter of the pipe (cm)

Equation 3.5 assumes that the contaminants instantaneously diffused across the entire pipe once they had permeated the pipe. Therefore, the estimated values using equation 3.5 are essentially the average pipe water concentration over the period of stagnation. To further simplify the calculation, the ratio of L_C to L_T was arbitrarily set to four values (1:1, 1:5, 1:10 and 1:20), and the period of stagnation, t, was set to 8 hours. With these simplifying conditions, the concentrations of benzene and toluene in pipe water were estimated for bulk concentrations or soil pore water concentrations ranging from 0.1 to 100 mg/L (see Figure 3.5).

As shown in Figure 3.5, the estimated bulk concentrations, C_{bulk}, (at steady state permeation) that would result in exceeding the benzene MCL (5 μg/L) in a 1-inch SIDR 9 IPS PE pipe for a stagnation period of 8 hours were approximately 1 mg/L, 4 mg/L, 8 mg/L and 15 mg/L, corresponding to the ratios (L_C/L_T) of 1:1, 1:5, 1:10 and 1:20, respectively. For identical bulk concentrations, however, the concentration of toluene in pipe-water would not exceed its MCL (1,000 μg/L) except at a high bulk concentration of 100 mg/L, which is not usually encountered in the field. Prevention of potential risk of benzene in drinking water should be a priority if PE pipes are exposed to gasoline-contaminated groundwater. Conservatively, if the bulk concentration of benzene is 1 mg/L or above, PE pipes must be rapidly replaced by other pipe materials (such as copper pipe) since the MCL in pipe-water will be exceeded within several hours. It should also be emphasized that Figure 3.5 was based on an 8-hour stagnation period. Under long-term stagnation (such as weeks), much lower bulk concentration will result in exceeding the MCL. For example, the concentration of benzene in pipe water was estimated to be 7 μg/L when the pipe was exposed to a bulk concentration of 50 μg/L for a stagnation period of 2 weeks at a ratio of 1:1.

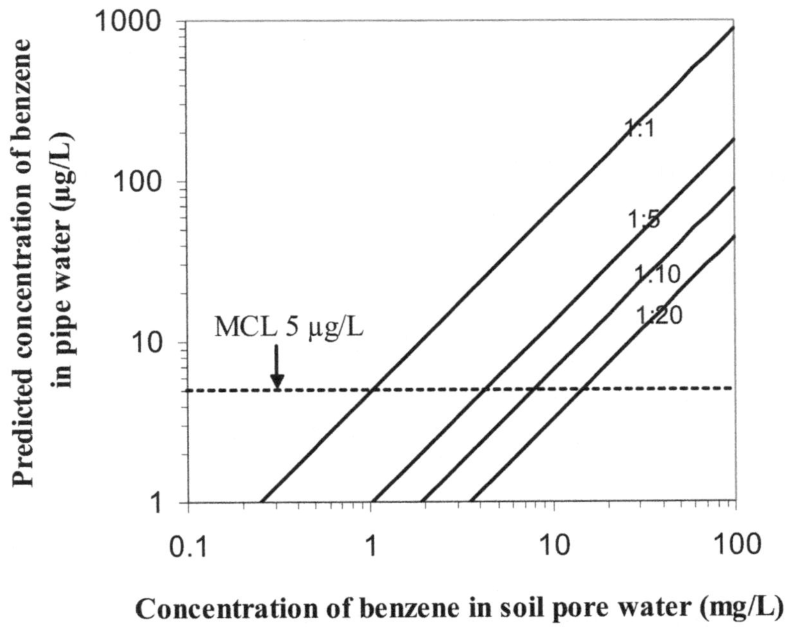

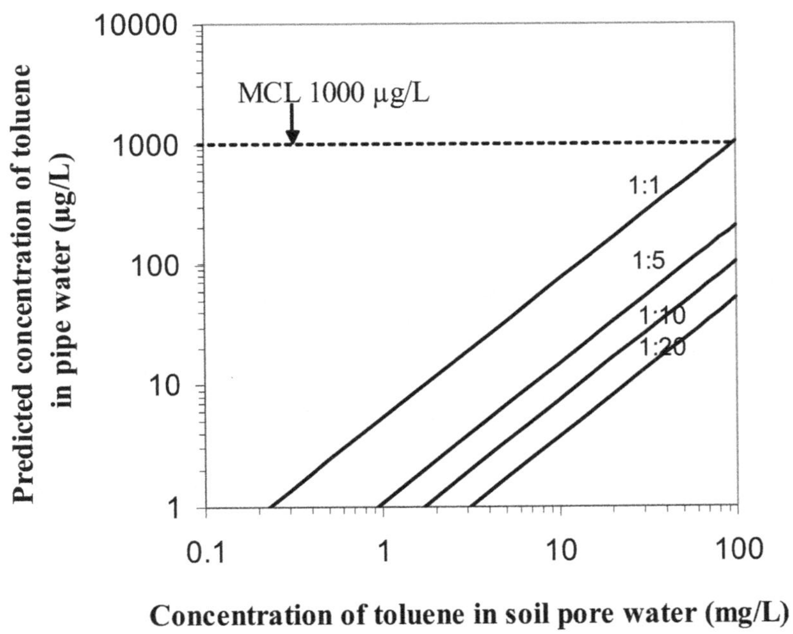

Figure 3.5 Concentrations of benzene and toluene in pipe water for various water bulk concentrations after 8-hour stagnation for 1-inch SIDR 9 PE pipe (ratios L_C/L_T = length of contaminated pipe/total length of pipe)

Under normal conditions, the contaminant concentration will depend on the water flow rate in the pipe. Assuming a full pipe flow and complete mixing of permeated contaminant in the pipe-water, the concentration of contaminant in pipe-water (C_{pw}) can be estimated by:

$$C_{pw} = \frac{M}{V} = \frac{P_m \times \pi \times O.D \times L_C \times t}{q \times t} = \frac{4 \times f(C_{bulk}) \times O.D. \times L_C}{I.D.^2 \times v} \tag{3.6}$$

where q = water flow rate (cm^3/day)
v = average water flow velocity (cm/day)

Other nomenclatures were defined previously. Typically, the average water flow velocity for plastic pipes, v, ranges from 0.06 m/s to 3.7 m/s (0.2 ft/s to 12 ft/s) (Chevron Phillips Chemical Company LP 2003, Uni-Bell 2001). By setting the average water flow velocities at six values (0.06, 0.30, 0.61, 1.52, 2.44, and 3.7 m/s (0.2, 1, 2, 5, 8, and 12 ft/s)), the length of contaminated pipelines, L_C, required (at steady state permeation) to exceed the MCL of benzene was estimated for a specific bulk concentration. The results are shown in Figure 3.6.

Figure 3.6 indicates that only very low water flow velocities (< 0.3 m/s (< 1 ft/s)) will pose a risk of exceeding the MCL for benzene under the conditions of full pipe flow. Given a water flow velocity of 0.3 m/s (1 ft/s) or above, a long contaminated pipeline (> 100 m) and a high bulk concentration (> 40 mg/L), which is not usually encountered in the real permeation incidents, would be needed to exceed the MCL. When the bulk concentration is below 10 mg/L and the length of contaminated pipe was less than 150 m, it is expected that the concentration of

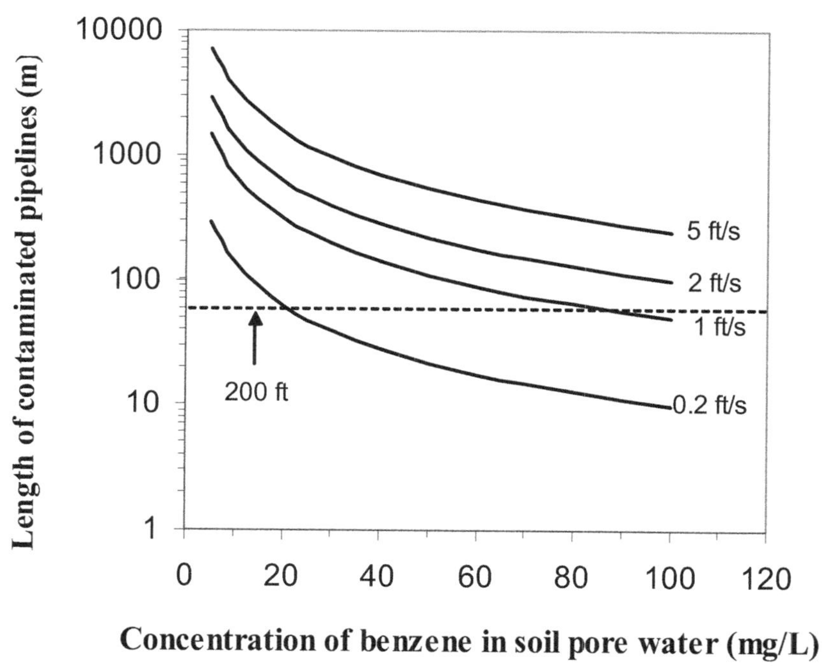

Figure 3.6 Length of contaminated pipe required to exceed the MCL of benzene (5 µg/L) in pipe water for various bulk concentrations and water flow velocities for 1-inch SIDR 9 PE pipe (L_C/L_T = 1:1)

benzene in pipe-water would be below its MCL for any typical water flow velocity. Due to the effect of dilution, the hydrocarbon concentrations in PE pipes with water flow would be less than those for conditions of stagnation.

The discussion above has been focused on 1-inch SIDR 9 IPS PE pipe. A key issue is to apply the data to predict the permeation behavior of contaminants for pipes with other dimensions. In equation 3.5, the inside diameter (I.D.) and outside diameter (O.D.) are known for a specific pipe size, but the steady state permeation rate (P_m) needs to be corrected for the effects of dimension. Based on the physicochemical theory of permeation (Crank and Park 1968), P_m can be expressed as:

$$P_m = C_0 \times \frac{P_C}{l} = C_0 \times \frac{D \times S}{l} \quad (3.7)$$

where C_0 = bulk concentration of contaminant surrounding the pipe (μg/L)
P_C = permeation coefficient (cm^2/s)
D = diffusion coefficient, (cm^2/s)
S = solubility (dimensionless)
l = wall thickness of pipe (cm)

S describes the partition of contaminant between bulk solution and polymeric material and in principle is a constant for a given contaminant-polymer pair. There are disagreements with respect to whether *D* is dependent on the polymer thickness. Generally, it is believed that *D* is independent of polymer thickness since *D* is only determined by the chemical characteristics of the polymer, the concentration (activity) of the contaminant compound, and the interactions between the contaminant and the polymer. However, Park et al. (1996) investigated the effects of thickness on diffusion coefficients for HDPE geomembrane exposed to four organic compounds (i.e., methylene chloride, toluene, trichloroethylene and m-xylene), and found that the diffusion coefficients decreased by 28 to 36% when the thickness increased from 0.76 mm to 2.54 mm at an initial concentration of 100 mg/L. Here, *D* is assumed to be a constant for various pipe sizes since this assumption would be conservative, but safe, to predict the permeation risk for large pipe dimensions. For different pipe sizes, therefore, P_m [$\mu g/cm^2$/day] is only a function of wall thickness of pipe (l). The steady state permeation rate for a specific pipe dimension can be deduced:

$$P_{mp} = P_{ms} \frac{l_s}{l_p} = \gamma P_{ms} \quad (3.8)$$

where P_{mp} = predicted steady state permeation rate for the wall thickness of l_p
P_{ms} = known steady state permeation rate for the wall thickness of l_s
γ = correction factor

Table 3.3 summarizes the values of γ for SIDR 7, SIDR 9, and DIPS DR 17.0 PSI 100 series of PE pipes. With the known correction factor, inside diameter, and outside diameter, the corresponding concentration of benzene in pipe-water was estimated after 8-hour stagnation for various bulk concentrations using equation 3.5 (see Figure 3.7).

Table 3.3
Correction factors (γ) for SIDR 7, SIDR 9, and DIPS DR 17.0 PSI 100 series of PE pipe to estimate the steady state permeation rates of contaminants

Pipe (inch)		O.D. (inch)	I.D. (inch)	Average wall thickness (inch)	O.D. (cm)	I.D. (cm)	Average wall thickness (cm)	Correction factor
SIDR 7	¾	1.070	0.824	0.123	2.718	2.093	0.312	0.992
	1	1.359	1.049	0.155	3.452	2.664	0.394	0.787
	1¼	1.785	1.380	0.202	4.534	3.505	0.514	0.604
	1½	2.086	1.610	0.238	5.298	4.089	0.604	0.513
	2	2.675	2.067	0.304	6.795	5.250	0.772	0.401
SIDR 9	½	0.760	0.622	0.069	1.930	1.580	0.175	1.769
	¾	1.018	0.824	0.097	2.586	2.093	0.246	1.258
	1	1.293	1.049	0.122	3.284	2.664	0.310	1.000
	1½	1.918	1.610	0.154	4.872	4.089	0.391	0.793
	2	2.537	2.067	0.235	6.444	5.250	0.597	0.519
DIPS DR 17.0 PSI 100	4	4.800	4.202	0.598	12.192	10.673	1.519	0.204
	6	6.900	6.039	0.861	17.526	15.339	2.187	0.142
	8	9.050	7.922	1.128	22.987	20.122	2.865	0.108
	10	11.100	9.716	1.384	28.194	24.679	3.515	0.088
	12	13.200	11.555	1.645	33.528	29.350	4.178	0.074
	14	15.300	13.392	1.908	38.862	34.016	4.846	0.064
	16	17.400	15.229	2.171	44.196	38.682	5.514	0.056
	18	19.500	17.068	2.432	49.530	43.353	6.177	0.050
	20	21.600	18.905	2.695	54.864	48.019	6.845	0.045
	24	25.800	22.582	3.218	65.532	57.358	8.174	0.038

Figure 3.7 indicates that small size pipes are more vulnerable to permeation than large size pipes. For example, approximately 400 μg/L of benzene in soil pore water would result in the exceedance of the benzene MCL for ½ inch SIDR 9 series pipe while the bulk concentration required to exceed the MCL for 2 inch SIDR 9 pipe would be 3.5 mg/L or higher. The higher susceptibility of small size pipes to permeation is supported by the findings of real permeation incidents, as reported in the study by Holsen et al. (1991a), where all permeation incidents for service lines were associated with 1 inch or less diameter pipes. For the same bulk concentration, the concentration of contaminant in pipe-water decreased with an increase in pipe size, mainly due to the increase of wall thickness and consequently the decrease of permeation rate.

In comparison with PE service lines, PE water mains have much lower susceptibility to permeation, particularly under a heavy contamination conditions. As shown in Figure 3.7, a bulk concentration of 10 mg/L would result in exceeding the benzene MCL in SIDR 7 and SIDR 9 series of PE service lines for a stagnation period of 8 hours, while that would not be expected to occur in DIPS DR 17.0 100PSI series of PE water mains. For water mains with a size of 14 inch or larger, the concentration of benzene in pipe-water would not exceed its MCL for a stagnation period of 8 hours even at an extremely high bulk concentration of 100 mg/L.

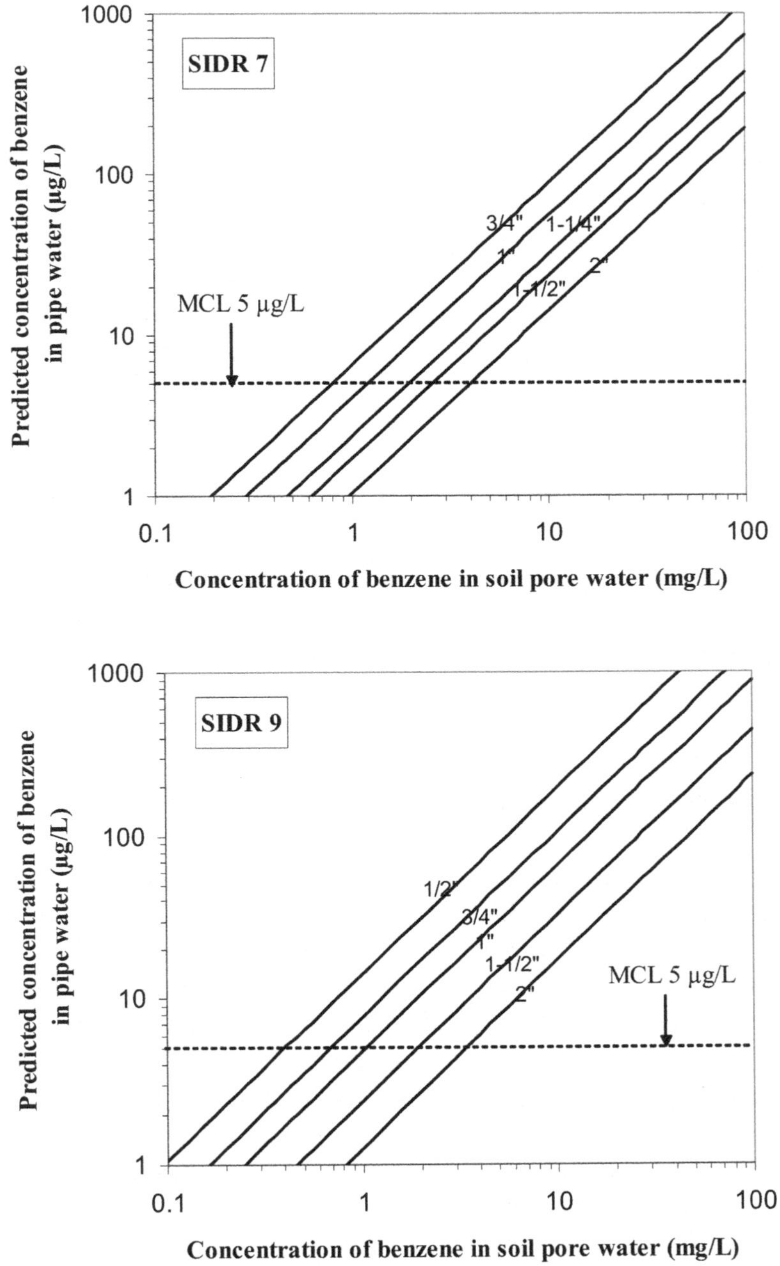

Figure 3.7 Predicted concentrations of benzene in pipe water for various bulk concentrations after 8-h stagnation for SIDR 7, SIDR 9 and DIPS DR 17.0 PSI 100 series of PE pipe (L_C/L_T = 1:1)

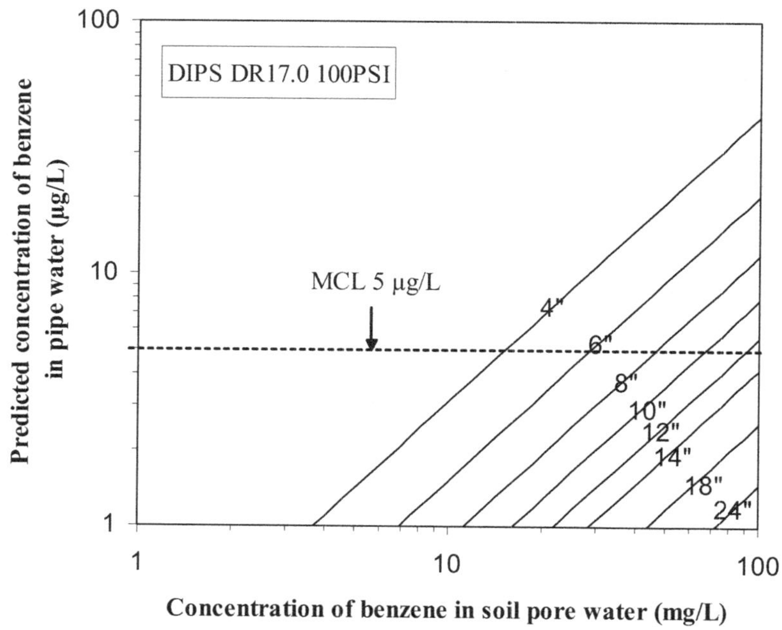

Figure 3.7 (Cont'd.) Predicted concentrations of benzene in pipe water for various bulk concentrations after 8-h stagnation for SIDR 7, SIDR 9 and DIPS DR 17.0 PSI 100 series of PE pipe (L_C/L_T = 1:1)

Exposure of PE Pipes to Unsaturated Gasoline-Contaminated Soil

The cumulative mass permeated per unit area for BTEX compounds at four levels of contamination are plotted in Figure 3.8. Toluene and xylenes were the major compounds detected in the pipe-water while the permeation of benzene through pipes was insignificant. These observations were consistent with the results found in the permeation incidents (Holsen et al. 1991a). In three reported incidents involving gasoline, toluene permeated to the greatest degree, followed by xylenes, ethylbenzne, and benzene (in the same order as shown in Figure 3.8a) while in the other cases associated with gasoline, xylenes permeated more readily, followed by toluene and ethylbenzene (in the same order shown in Figure 3.8b and c). Holsen et al. (1991a) attributed the low permeation of benzene to its relatively higher polarity. The properties of polarity, however, cannot explain the higher and faster permeation of benzene than the other BTEX compounds when pipes were exposed to free gasoline or gasoline-contaminated water. One possible reason, as indicated in this study, is the relatively low benzene concentration remaining in the soils, due to the high soil sorption affinity of benzene among the BTEX compounds (Zytner 1994). In the gasoline-contaminated groundwater, however, a relatively high concentration of benzene is generally detected because of its high solubility in water (Cline et al. 1991). From the perspective of permeation risk, therefore, the plastic pipes buried in the water-saturated zone will be more easily permeated by benzene than in the vadose zone. The case of benzene indicates that knowledge of environmental fate and transport is critical to evaluate the potential permeation risk of contaminants since the distribution of chemicals among environmental media determines the contaminant levels plastic pipes might encounter in the field.

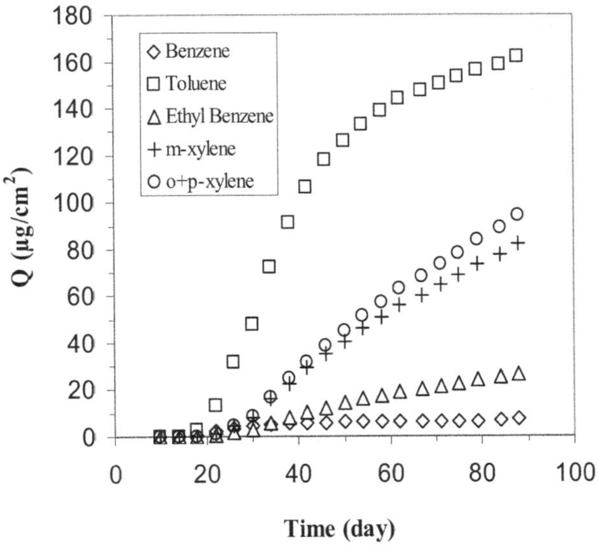

(a) Initial BTEX concentration (mg/kg dry soil): total BTEX, 1216; benzene, 72; toluene, 531; ethylbenzene, 85; m-xylene, 267; o+p-xylene, 261.

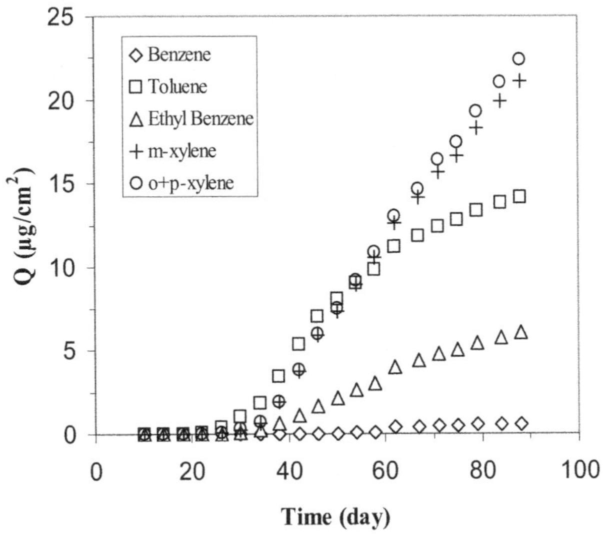

(b) Initial BTEX concentration (mg/kg dry soil): total BTEX, 388; benzene, 3; toluene, 119; ethylbenzene, 36; m-xylene, 115; o+p-xylene, 115.

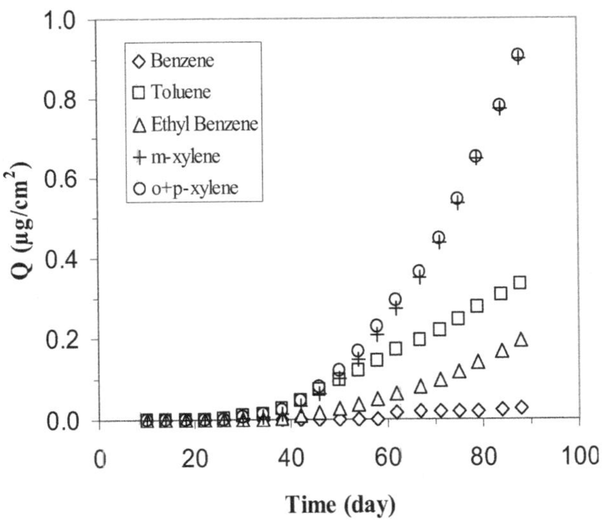

(c) Initial BTEX concentration (mg/kg dry soil): total BTEX, 89; benzene, <1; toluene, 9; ethylbenzene, 9; m-xylene, 34; o+p-xylene, 37.

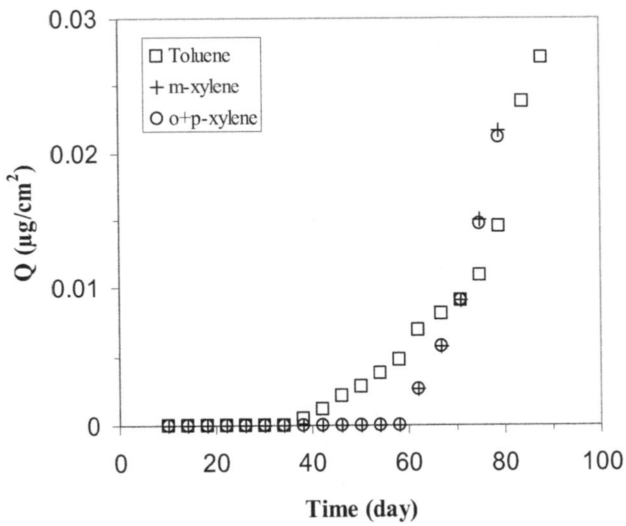

(d) Initial BTEX concentration (mg/kg dry soil): total BTEX, 32; benzene, <1; toluene, 3; ethylbenzene, 3; m-xylene, 12; o+p-xylene, 14.

Figure 3.8 Cumulative mass permeated per unit area for BTEX compounds in PE pipes exposed to gasoline-contaminated soils

In the unsaturated soil, plastic pipe permeation was believed to occur from vapor phase (Holsen et al. 1991b, Park et al. 1991). This study did not collect the vapor samples for BTEX analysis and thus the exact concentration of BTEX in soil gas was unknown over the experimental period. At the end of the experiments, soil samples were taken from different positions (top, middle and bottom) of the glass bottle and the BTEX concentrations in the soils were found to be relatively uniform. However, for each of the four contamination levels, more than 90 percent of total BTEX in soils was lost over the entire experimental period of 3 months. The loss of toluene was relatively higher than that of xylene. The losses were mainly due to the sorption by the pipe and volatilization. Generally, the diffusion coefficients of organic compounds in the vapor phase through soil pores are several orders of magnitude larger than the diffusion coefficients through plastic pipes (Holsen et al. 1991b). Therefore, the mass transfer in soils may be rapid enough to hinder the development of a concentration gradient in the soils. The significant drop of BTEX concentration in soils indicated that the pipes were exposed to variable contaminant levels, which resulted in somewhat less than ideal permeation curves for constant exposure concentration. As shown in Figures 3.8a and b (initial BTEX concentrations of 1,216 and 388 mg/kg), the permeation curves, especially for toluene, flattened out at later time periods. However, Figures 3.8c and d still showed a linear increase of cumulative mass permeated per unit area with time after steady state permeation was reached – a typical curve predicted by permeation theory.

The time lag method approach cannot be used to estimate the diffusion coefficients of BTEX for this case since the experimental conditions did not meet the requirement of "constant exposure concentration." However, it may be reasonable to assume that the exposure concentrations did not significantly decrease at the early time period and thus it is still possible to apply the time lag method to calculate the diffusion coefficients based on the early breakthrough data. The first 50 days of data were used to estimate the diffusion coefficients for the individual BTEX compounds as shown in Figures 3.8a and b. The permeation parameters of toluene, ethylbenzene and xylene under the test conditions are summarized in Table 3.4.

Generally, the diffusion coefficients (D) in Table 3.4 were of the same order of magnitude as those found in the aqueous experiments (10^{-9} cm^2/s), were concentration-dependent, and showed signs of synergistic effects. The synergistic effect was demonstrated by the diffusion coefficients of the three xylene isomers. In a previous study (Saleem et al. 1989), significant differences in the diffusion coefficients were detected for the three xylene isomers due to the effect of molecule structure (shape) on diffusion. In this study, however, the three compounds shared identical permeation characteristics when they coexisted in the soils. There are no means of comparing the results of diffusion coefficients found in this study as none of previous experimental studies have specifically investigated the permeation of organic contaminants through PE pipe buried in the unsaturated soil. In addition, comparing the diffusion coefficients obtained from the three permeation tests (free product gasoline, gasoline-contaminated water, and gasoline-contaminated unsaturated soil) based on the values of activity is a challenge due to the lack of consistent definitions of activity (Park et al. 1991).

Table 3.4
Permeation parameters for toluene, ethylbenzene and xylenes in PE pipes exposed to gasoline-contaminated soils

Chemical	External bulk concentration (mg/kg dry soil)	Breakthrough time (day)	Time lag (day)	Steady state permeation rate ($\mu g/cm^2/day$)	Diffusion coefficient ($\times 10^{-9}$ cm^2/day)
Toluene	531	10	20	5.1	9.3
	119	14	30	0.43	6.1
	9	26	34	0.0061	5.4
	3	42	69	0.0014	2.7
ethylbenzene	85	18	22	0.5	8.4
	36	22	32	0.12	5.8
	9	38	53	0.0054	3.5
	3	-	-	-	-
m-xylene	267	14	23	1.5	8.1
	115	18	33	0.43	5.7
	34	30	55	0.027	3.4
	12	62	65	0.0016	2.9
o+p-xylene	261	14	23	1.7	8.1
	115	18	33	0.43	5.7
	37	30	55	0.027	3.4
	14	62	66	0.0016	2.8

The steady state permeation rates versus the bulk (external) concentrations for toluene and total xylenes in unsaturated soils are plotted in Figure 3.9. Two empirical correlation equations as shown below were obtained for the steady state permeation rate, P_m [$\mu g/cm^2/day$)] and the bulk concentration, C_{bulk} (mg/kg dry soil):

$$P_m = 2 \times 10^{-4} C_{bulk}^{1.6054} \quad \textit{for toluene}$$
$$P_m = 3 \times 10^{-6} C_{bulk}^{2.2594} \quad \textit{for total xylenes} \qquad (3.9)$$

The above equations were used to predict the concentrations of toluene and total xylenes in the pipe-water under various bulk concentrations after 8 hours, 1 day, 7 days and 30 days of stagnation (Figure 3.10). The estimation approach has been described in the previous sections. To simplify the calculation, the ratio of L_C to L_T in equation 3.6 was set to 1. As shown in Figure 3.10, the estimated bulk concentration, C_{bulk}, that would result in exceeding the MCL for toluene (1,000 µg/L) in a 1 inch SIDR 9 IPS PE pipe were approximately 300, 130, 50 and 15 mg/kg dry soil, corresponding to the stagnation period of 8 hours, 1 day, 7 days and 30 days, respectively. For total xylenes, the bulk concentration needed to exceed the MCL (10 mg/L) were 1,000, 600, 250, 130 mg/kg dry soil for stagnation periods of 8 hours, 1 day, 7 days and 30 days, respectively. Efforts were also made to predict the concentration of toluene and total xylenes in the pipe water under the condition of continuous water flow. Due to relatively high MCLs of toluene and total xylenes, exceeding the MCLs would be unlikely under typical flow velocities. Using the correction factors presented in Table 3.3, the concentrations of toluene in pipe-water for various soil bulk concentrations was estimated as shown in Figure 3.11 for SIDR 7 and SIDR 9 series of PE pipes after 7-day stagnation.

Caution should be exercised when utilizing the results of this study to predict the permeation behavior of contaminants through plastic pipes in unsaturated soil because other environmental factors may have potential impact on the permeation process. Soil characteristics such as porosity and water moisture have an effect on the transport of contaminants in the porous media and thus possibly affect the availability of these compounds for permeation through pipe materials. The permeation process is temperature dependent. It has been established (Naylor 1989) that for small temperature changes between 5 to 25° C, the Arrhenius relationship can be used to describe the dependence of the diffusion coefficients on temperature as follows:

$$D = D_0 e^{\left(-\frac{E_d}{RT}\right)} \tag{3.10}$$

where E_d = activation energy of diffusion (cal/mole)
D_o = diffusion coefficient at 25° C (cm^2/s)
D = diffusion coefficient at other temperatures (cm^2/s)
R = ideal gas constant (cal/mole °K)
T = absolute temperature (°K)

Based on the above, the diffusion coefficients are expected to be lower at lower temperatures. The experiments were conducted at room temperature (23° C) while the average soil temperature in the field may be in the range of 14 – 15° C. Therefore, it is likely that the measurements at room temperature will overestimate the permeation risk in the field.

Impact of Soil Organic Matter on Permeation (Saturated Soils Experiments)

Sorption of BTEX into the soil organic matter is rapid, as shown in Figure 3.12. BTEX in the soil pore water for the organic topsoil and sand-topsoil mixture dropped dramatically within the first week (mainly due to sorption to the soil) and then decreased slowly (mainly due to sorption to the pipe and volatilization losses). Under otherwise identical initial conditions, the BTEX concentrations in soil pore water after 55 days were lowest in the organic topsoil (approx. 30 mg/L as compared to an initial concentration of approximately 128 mg/L) which had the highest organic carbon content (5.1%). The BTEX concentrations in the soil pore water of the sand-topsoil mixture with an organic carbon content of 1.9 % were approximately 50 mg/L. The BTEX concentrations in soil pore water for the silica sand with insignificant organic carbon content were approximately 65 mg/L.

The different BTEX concentrations were attributed to different BTEX sorption capacities for different types of soil. Generally, soils behaved as dual sorbents: the mineral matter functions as a conventional adsorbent (adsorption), and the soil organic matter as a partition medium rather than a conventional solid adsorbent (partition) (Chiou and Peters 1981). Earlier studies found that water and organic chemicals compete for sorption sites on mineral surfaces and a relatively high humidity may significantly inhibit the mineral adsorption of organic vapor. Since soils were saturated with water in this study, partitioning into soil organic matter is probably the dominant mechanism of BTEX uptake from water. The higher organic matter in the organic topsoil resulted in greater soil uptake of BTEX and a significant decrease of BTEX concentrations in the soil pore water.

As demonstrated in the section on "Exposure of PE Pipes to Gasoline-Contaminated Groundwater", the permeation rates of BTEX through PE pipes were strongly dependent on the external bulk concentration. The sorption of BTEX by soils decreased the BTEX concentration in the soil pore water and, thereby, decreased the permeation rate of BTEX through PE pipes (see Figure 3.13). The breakthrough times of benzene for silica sand, sand-topsoil mixture, and organic topsoil were 13 days, 16 days, and 19 days, respectively. Since the external bulk

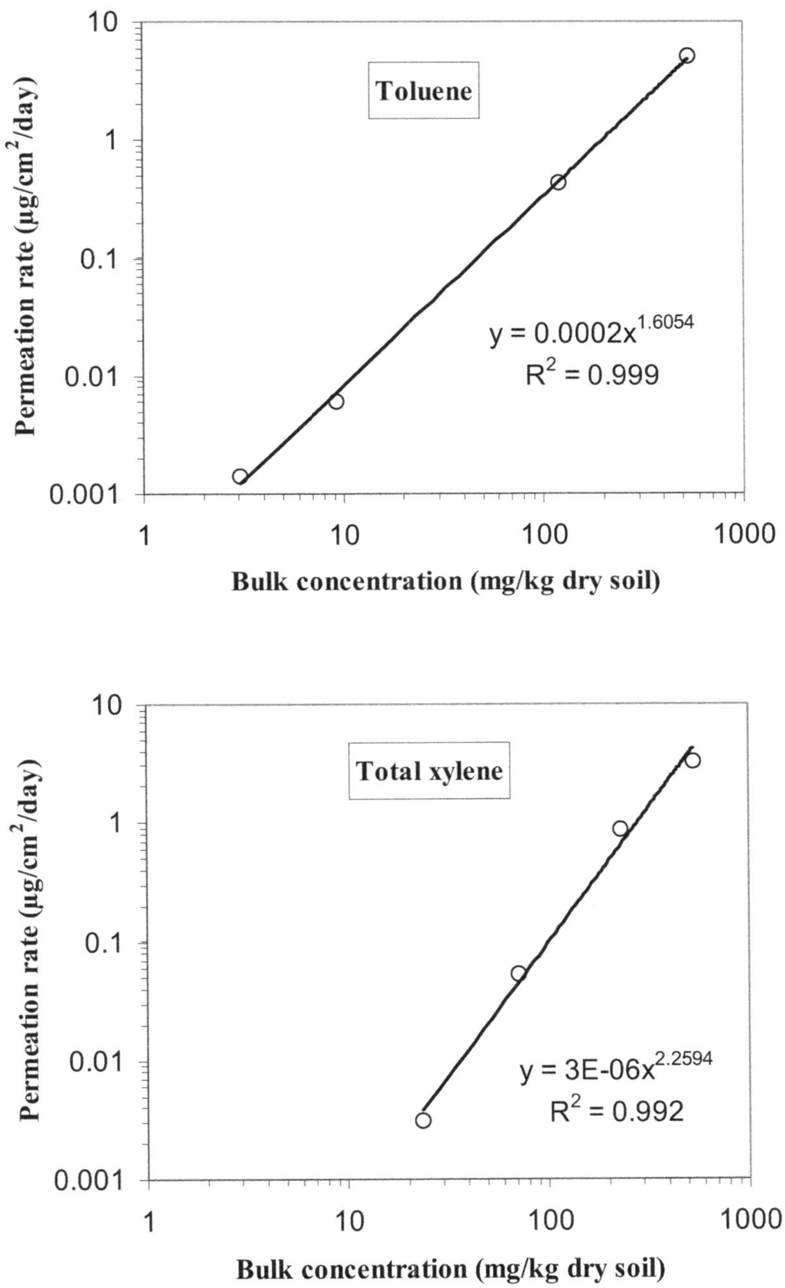

Figure 3.9 Correlation between the steady state permeation rates and the external bulk concentrations in soil for toluene and total xylene

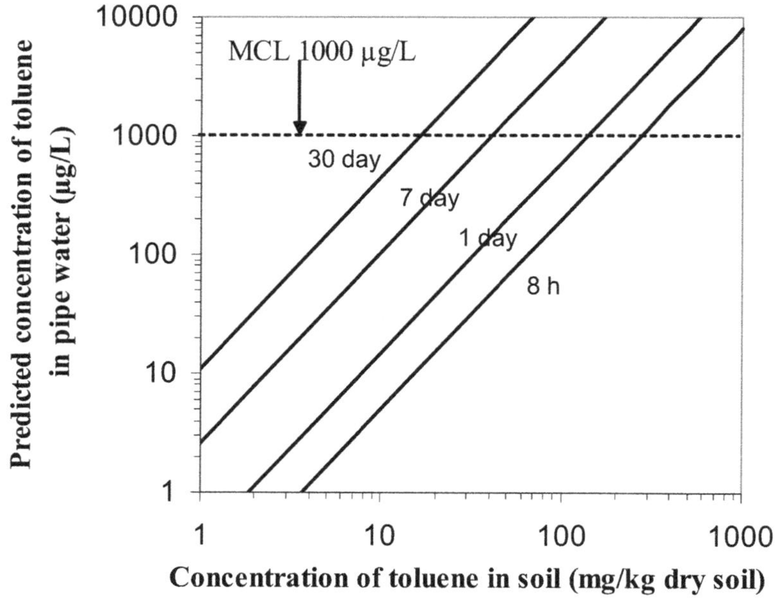

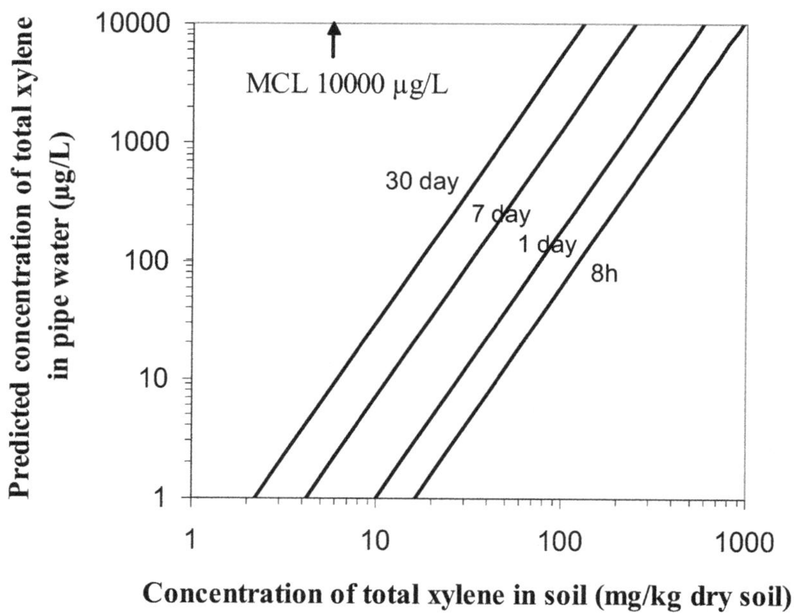

Figure 3.10 Predicted concentrations of toluene and total xylenes in pipe water for various soil bulk concentration after a period of stagnation for 1-inch SIDR 9 PE pipe (L_C/L_T = 1:1)

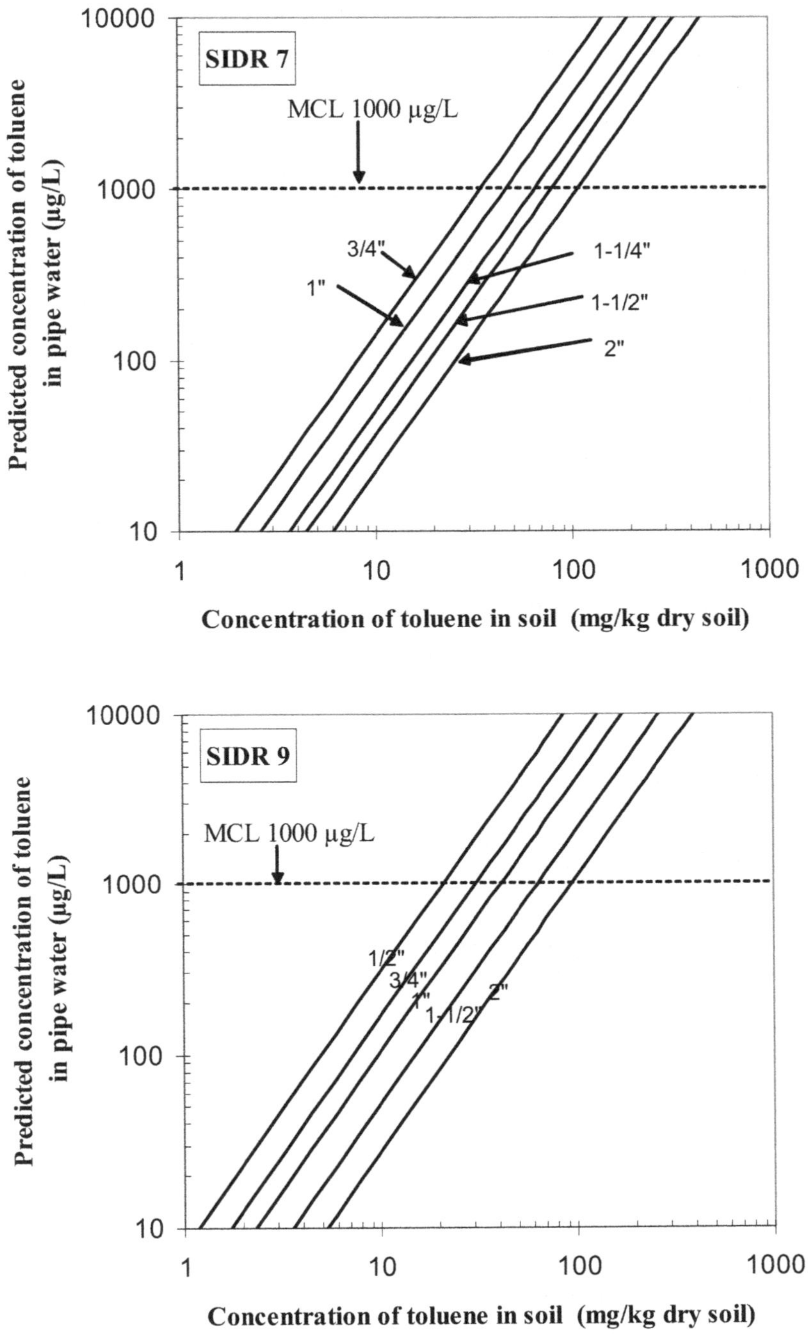

Figure 3.11 Predicted concentrations of toluene in pipe water for various soil bulk concentrations after 7-day stagnation for SIDR 7 and SIDR 9 series of PE pipe (L_C/L_T = 1:1)

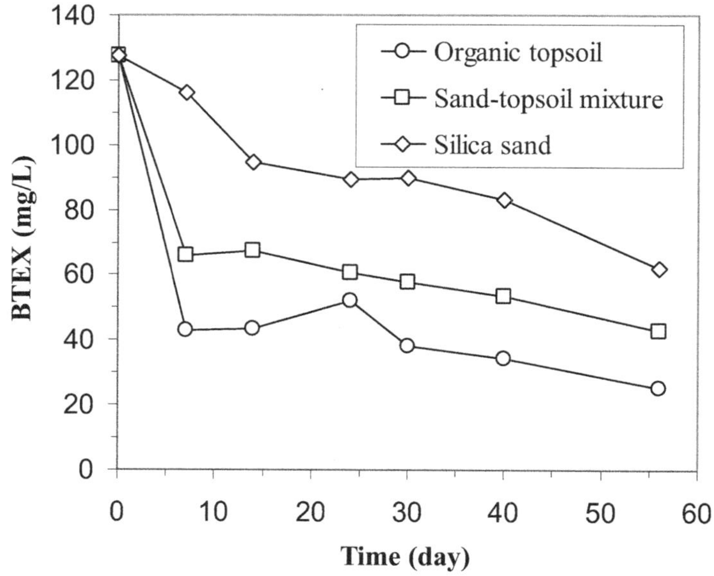

Figure 3.12 Change of total BTEX concentrations in soil pore water with exposure time

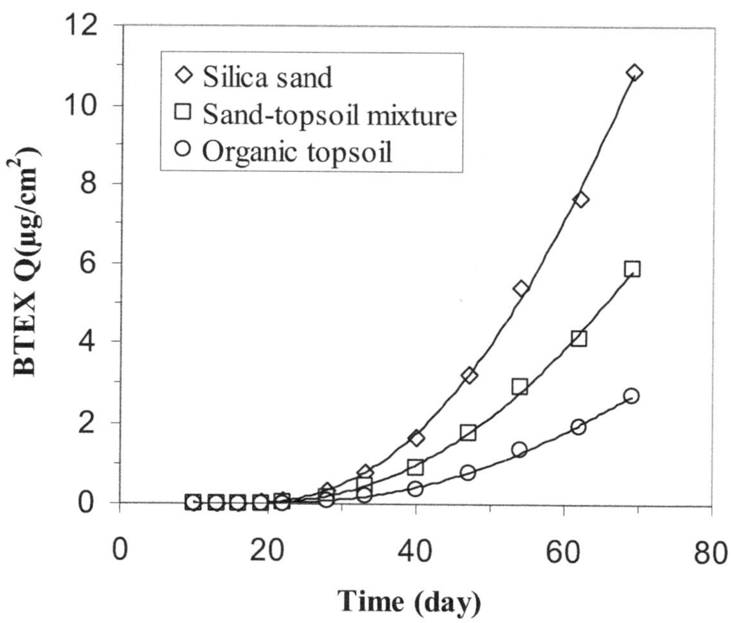

Figure 3.13 Cumulative mass of total BTEX permeated per unit area in PE pipes

concentration varied with exposure time, steady state permeation was not reached and, therefore, the diffusion coefficients of BTEX for the three different types of soils could not be estimated. The relative permeation rates, however, can be compared with each other. For example, after two months of exposure, the cumulative mass of BTEX permeated per unit area through the pipe buried in silica sand was nearly twice that of the sand-topsoil mixture, and four times that of organic topsoil. However, high soil organic matter can not be relied upon to protect pipes from permeation because such soils will eventually reach their maximum adsorption capacities under field conditions. The effects of soil organic matter on permeation found in this study were consistent with the results of the study of Holsen et al. (1991b).

CONCLUSIONS

Pipe-bottle tests were conducted to investigate the permeation of petroleum-based aromatic compounds (BTEX) through 1-inch SIDR 9 PE pipe under simulated field conditions, including gasoline leaks and spills, gasoline-contaminated groundwater, and gasoline-contaminated soil with various levels of contamination. Based on the analysis of experimental results, the following conclusions can be drawn:

(i) PE pipe was rapidly permeated when exposed to either free product gasoline, gasoline-contaminated water solutions, or gasoline-contaminated unsaturated soils. Benzene and toluene accounted for nearly all the total BTEX that permeated into the pipe-water when pipes were buried in the silica sand saturated with free product gasoline or gasoline-contaminated water solutions. Toluene and xylenes were the major compounds detected in the pipe-water when pipes were buried in the gasoline-contaminated unsaturated soils, while permeation of benzene and ethylbenzene was insignificant.

(ii) Using the time-lag method, the diffusion coefficients of BTEX compounds were estimated to be on the order of 10^{-8} cm^2/s when exposed to free product gasoline and on the order of 10^{-9} cm^2/s when exposed to gasoline-contaminated water solutions or unsaturated soils. The diffusion coefficients were concentration-dependent and might be influenced by the synergistic effect of organic mixtures.

(iii) The steady state permeation rates of BTEX compounds were strongly dependent on the bulk concentration. Empirical correlation equations (Equation 3.4 and 3.9) were derived for the permeation rates and bulk concentrations. The equations can be used in conjunction with equation 3.5 and 3.6 to predict the concentration of contaminants in pipe-water for PE pipes exposed to various bulk concentrations after a period of stagnation or under continuous flow conditions. Predictions using the empirical equations indicated that small size pipes were more vulnerable to permeation than large size pipes, and pipes with water stagnation periods posed a much higher risk of exceeding the MCLs of BTEX compounds relative to pipes with continuous water flow. The information needed to use the equations includes contaminant concentrations in groundwater or in unsaturated soils, pipe dimensions (inside and outside diameter), length of stagnation periods, and the water flow velocity of the pipe.

(iv) A PE pipe buried in soils of high organic matter was permeated to a lesser extent than a pipe buried in a soil of low organic matter. This is due to the sorption of organic compounds by the organic matter resulting in a lower soil pore water concentration than soils with low organic matter.

CHAPTER 4
POLYVINYL CHLORIDE (PVC) PIPES - LABORATORY STUDIES UNDER SIMULATED ENVIRONMENTAL CONDITIONS

OVERVIEW

The use of polyvinyl chloride (PVC) pipes in water distribution systems has increased dramatically within the past several decades, mainly due to their excellent abrasion, impact, and corrosion resistance (Uni-Bell 2001). However, PVC pipes buried in contaminated sites may be vulnerable to permeation by contaminants external to the pipe through the porous, non-metallic materials into the drinking water (US EPA, 2001). According to a survey completed in the 1980s (Thompson and Jenkins 1987), PVC pipes were involved in 15% of total incidents of drinking water contamination by permeation.

Some efforts such as the gravimetric sorption test (Berens 1985), the pipe-bottle exposure test (Vonk 1985), and pilot-scale exposure studies of PVC water mains (Olson et al. 1987) have been applied to ascertain the susceptibility of PVC pipes to permeation by organic chemicals. These studies indicated that: (i) permeation of organic compounds through PVC pipes was negligible for low concentrations of pollutants; (ii) at high concentrations of pollutants, such as in the case of a gross chemical spill, swelling of PVC by certain organic chemicals can cause rapid chemical penetration of PVC pipes; and (iii) gasketed PVC pipes showed faster permeation compared to ungasketed PVC pipes.

Previous studies have greatly improved the level of understanding of permeation risks. However, there is still not enough information regarding whether PVC pipe is an effective barrier against long term exposure to commercial gasoline, as in a typical contaminated site or gasoline spill (Holsen et al. 1991b). Moreover, most of the previous studies focused on organic solvents such as toluene, and these experiments may not be sufficient to reflect the comprehensive permeation behavior of pipes in the subsurface environment since permeation can occur either from the vapor phase, aqueous phase, or nonaqueous liquid phase (NAPL).

The aim of this study was to investigate the permeation of organic contaminants through PVC pipes under extremely aggressive environments, including neat organic solvents, free product gasoline, saturated organic vapors, as well as conditions that better reflect real field conditions, such as aqueous solutions of solvents and aqueous solutions saturated with gasoline. Three organic compounds, benzene, toluene and trichloroethylene (TCE) were selected for study on the basis of their occurrence in contaminated soil and groundwater. The results serve as a base for predicting the permeation behavior of organic contaminants through PVC pipe in a wide range of field conditions. The data obtained from this study were also used to evaluate the usefulness of predictive pipe materials tests (Chapter 5).

MATERIALS AND METHODS

Materials

One-inch PVC pipes (SCH 40 450 psi@73F ASTMD NSF-PW) were purchased new from local stores. Tests were not conducted on old, used, or compromised pipes since the age, exposure history, and conditions are typically not known. Benzene, toluene and trichloroethylene (TCE) solvents were used and were purchased from Fisher Scientific (Chicago, IL). Premium gasoline was purchased from local stores and the average concentrations of benzene, toluene, ethylbenzene, m-xylene, and o+p-xylene (BTEX) were determined using gas chromatography to be 19.8 g/L, 75.9 g/L, 14.7 g/L, 33.7 g/L, and 32.5 g/L, respectively. BTEX compounds were roughly 25% by volume of the gasoline – slightly higher than 19 – 24% for various commercial gasolines reported by Cline et al. (1991). Silica sand (Granusil 4030) was purchased from UNIMIN Corporation, Portage WI. The sand contained very low organic matter (≈ 0%) and had a consistent uniform particle size diameter of 0.25 mm.

Experimental Apparatus

The gasoline-exposure experiments were conducted in a pipe-bottle apparatus, consisting of a 10-L glass bottle with two PVC pipes mounted horizontally through holes drilled in the glass (Figure 4.1a). A chemically resistant sealant, Loctite® epoxy putty (Henkel Technologies, Rocky Hill, CT), was used to seal the connection between the glass bottle and the pipe. No leaks were found over a 2-year experimental period. The ends of the pipes were sealed with Teflon® plugs equipped with brass needle valve assemblies for filling and draining the pipes.

The solvent-exposure experiments were conducted in a "mini" pipe-bottle apparatus consisting of a 1-L glass bottle with one PVC pipe mounted horizontally through holes drilled in the glass (Figure 4.1b). The same sealant as described above was used to seal the connection between the glass bottle and the pipe. The ends of the pipes were sealed with Teflon® plugs. The Teflon® plug had a small hole that was plugged with a brass plug to allow filling and draining of the water inside the pipe with a glass syringe. The bottle was capped with a Teflon®-lined cap. All experiments were conducted at room temperature (23 ± 1.5° C) for reasons of operational convenience and economy.

Experimental Procedures

PVC Pipes Exposed to Pure Solvents

Three "mini" pipe-bottle apparatuses were set up with a PVC pipe directly exposed to pure benzene, toluene and TCE, respectively. In this test, the pipe segments were first filled with deionized water and the solvents were then added into the bottle until a visible liquid level of solvent appeared above the pipes. To minimize the amount of solvents added and thus reduce the amount of hazardous waste produced, the bottle space below the pipes was filled with glass beads. The pipe-water was sampled each day and analyzed for the presence of the target compounds. After each sampling event, the pipe was immediately flushed three times with deionized water before fresh deionized water was added and the pipes sealed.

PVC Pipes Exposed to Organic Vapors

Three "mini" pipe-bottle apparatuses were set up with a PVC pipe directly exposed to the vapor of benzene, toluene and TCE, respectively. In this test, the pipe segment was first filled with deionized water and 100 mL solvent was then transferred into the bottom of the bottle. The visible liquid level of the solvent was around 3-cm below the lowest edge of the pipe segment. Preliminary experiments found that organic vapor condensed on the pipe surface. To avoid the

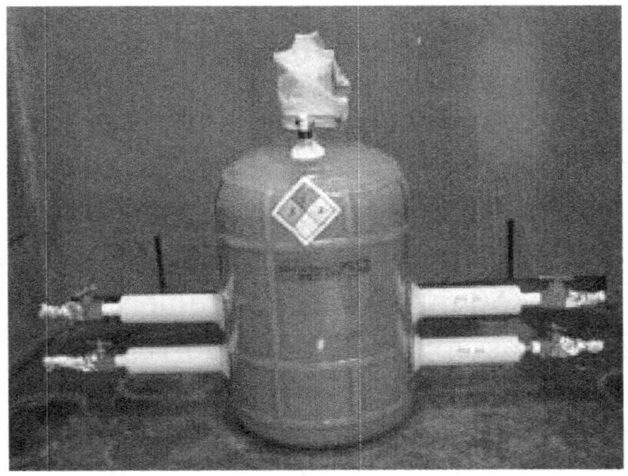

(a)

(b)

Figure 4.1 Pipe-bottle apparatuses (a) 10-L glass bottle (b) 1-L glass bottle

effect of condensation and to simulate the conditions closer to field conditions with humidity close to saturation, an opened glass vial containing 10 ml of deionized water was mounted between the pipe segment and the inner wall of the bottle. Pipe water was sampled at regular intervals and analyzed for the presence of the target compounds. After each sampling event, the pipe was immediately flushed three times with deionized water before fresh deionized water was added and the pipes sealed.

PVC Pipes Exposed to Water Saturated with Organic Solvents

Aqueous solvent-saturated solutions represent extreme field conditions encountered by PVC pipes in contaminated groundwater. To simulate such conditions, three "mini" pipe-bottle apparatuses were set up with a PVC pipe exposed to aqueous saturated solutions of benzene, toluene and TCE, respectively. To prepare the benzene or toluene-saturated solutions, a mixture of 200 mL solvent and 3.5 L deionized water was moderately stirred with a magnetic stirrer in a glass bottle for 48 hours and then allowed to sit undisturbed for 2 hours. To prepare the TCE-saturated solution, a mixture of 50 mL TCE with deionized water was vigorously stirred with a magnetic stirrer under zero head-space conditions for 48 hours in a 2-L volumetric flask with a Teflon® stopper, and then allowed to sit undisturbed for 2 hours. Preliminary experiments indicated that the above procedures produced reproducible saturated solutions. The measured concentrations of benzene, toluene and TCE were 1,767 mg/L, 501 mg/L and 1,093 mg/L, respectively, and were close to the solubilities reported in the literature.

For the permeation experiment, aqueous saturated solution was transferred into the pipe-bottle apparatus using a Masterflex pump with Teflon® tubing. To maintain the benzene and toluene concentrations at saturation, 50 mL benzene or toluene was added into the corresponding bottle so that the aqueous phase was covered by the solvent. The concentration of the TCE-saturated solution was maintained at saturation by keeping approximately 10 mL of TCE NAPL at the bottom of the bottle with gentle stirring. Headspace was minimized in all experiments. Pipe water samples were taken each day to estimate the breakthrough time.

PVC Pipes Exposed to Gasoline

In these experiments, pipe segments were filled with deionized water. The 10-L bottle was then filled with silica sand and premium gasoline until the 10-L bottle was nearly filled with silica sand and a visible liquid level of gasoline appeared above the surface of the sand. The liquid surface of solvent in the 10-L bottle was monitored visually and there was no significant drop in the liquid surface during the experimental period. Pipe water was removed each month and analyzed for the presence of the BTEX compounds. The duration of the experiment was 2 years.

PVC Pipes Exposed to Water Saturated with Gasoline

To prepare the gasoline-saturated water solution, a mixture of 1 L premium gasoline and 10 L deionized water was moderately stirred in a glass bottle for one week and then allowed to sit undisturbed for 3 days. The aqueous phase was then transferred into the pipe-bottle apparatus using a Masteflex pump with Teflon® tubing. To maintain a saturated solution, 1-L of fresh premium gasoline was added to the bottle so the aqueous phase was covered with gasoline. The headspace was also minimized. During the experimental period, the mixture of water and gasoline was routinely stirred and precautions were taken to avoid direct contact between pipe and gasoline. Pipe water was sampled at regular intervals and analyzed for the presence of the BTEX compounds. The duration of the experiment was 2 years.

Analytical Methods

BTEX or TCE in water samples were determined using a gas chromatograph (Tracor 540) equipped with a packed column (1.8 m (6 ft) × 2 mm; 1% SP1000 on 60/80 mesh Carbopack B), a photoionization detector, and an automated purge and trap concentrator (Tekmar LSC2/ALS). The detection limits for benzene, toluene, ethylbenzene, m-xylene o+p-xylene and TCE were 0.24, 0.24, 0.26, 0.29, 0.53, and 1.2 μg/L, respectively. All experiments were conducted at room temperature (23 ± 1.5°C).

RESULTS AND DISCUSSION

PVC Pipes Exposed to Pure Solvents

Significant permeation through PVC pipes occurred within 6.5 days for TCE, 16 days for toluene, and 20 days for benzene (Figure 4.2). No detectable permeation occurred through PVC pipes until a breakthrough point was reached. At that point, permeation occurred at a constant rate. The permeation curves for PVC pipes were consistent with those of Vonk (1985) who reported the permeation of a mixture of toluene and m-xylene through PVC pipes. At steady state permeation conditions, the concentrations of these compounds in the pipe-water were found to be close to their water solubilities at the end of the 1-day sampling intervals.

Permeation of solvents through PVC pipes is dependent on the interaction between organic solvents and pipe materials. In this study, the chlorinated organic solvent (TCE) was shown to permeate PVC pipes more rapidly than the aromatic compounds (benzene and toluene), which is in agreement with the findings of Vonk (1985). The time for complete breakthrough is also related to the composition of the pipe material and the pipe wall thickness. For example, the breakthrough time for toluene was variable for different studies - 5 days for PVC pipe with a wall thickness of 3.5 mm (Vonk, 1985), 16 days for PVC pipe with a wall thickness of 3.6 mm (this study), and nearly 40 days for commercial PVC water mains (Olson et al., 1987).

Significant swelling of pipe materials occurred as a result of the permeation process. All pipe specimens showed signs of softening and localized ballooning. To assess whether the PVC pipe would be restored to its original properties when the external contaminant source was removed, the organic solvent in the bottle was emptied and the pipes allowed to stand in the fume hood for about half a year (for the benzene and toluene experiments, only). During this time, the pipe segments were filled with deionized water and drained regularly. On examination after 6 months, the rubberization of the PVC pipes continued to persist and relatively high concentrations of organic compounds were detected in the pipe-water. These results indicated that once PVC pipes have been permeated, the permeation cannot be readily corrected by removing the contaminated soil and flushing the pipe line with clean water for a protracted period of time.

PVC Pipes Exposed to Organic Vapors

As shown in Figure 4.3, the permeation curves obtained in the vapor experiments were very similar to those obtained in the liquid solvent experiments. TCE permeated first, followed by toluene and benzene. For a specific organic compound, once breakthrough occurred, the permeation rate in the organic vapor was nearly identical to that in the neat solvent, as demonstrated by an identical slope of the cumulative mass permeated per unit area versus the exposure time. However, there was a delay in breakthrough under exposure to organic vapors. The breakthrough times under exposure to TCE vapor, toluene vapor and benzene vapor were estimated to be 13 days, 28 days and 31 days, respectively. Such delay is probably due to the effect of mass transfer which will be discussed later when comparing the data obtained from solvent, vapor and aqueous experiments.

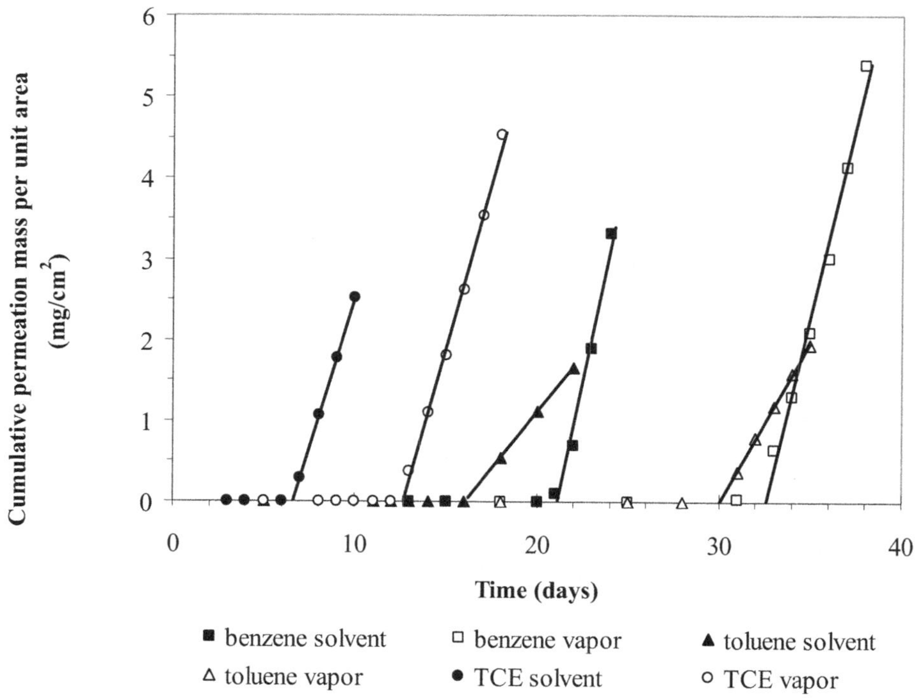

Figure 4.2 Cumulative mass permeated per unit area for PVC pipes exposed to benzene, toluene, and trichloroethylene

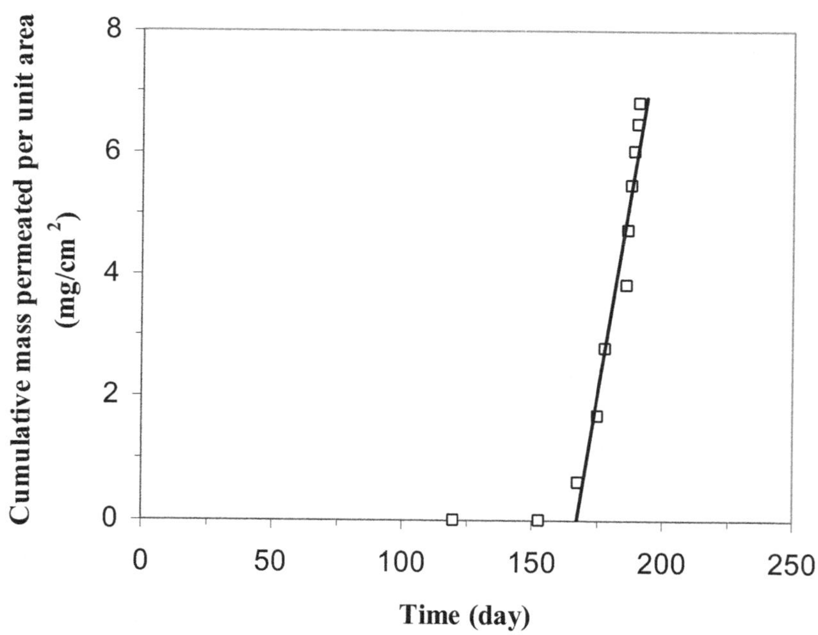

Figure 4.3 Cumulative mass of TCE permeated per unit area for 1-inch PVC pipe exposed to TCE-saturated aqueous solution

PVC Pipes Exposed to Water Saturated with Organic Solvents

The permeation of TCE and benzene through PVC pipe in saturated aqueous solutions is shown in Figure 4.3 and Figure 4.4, respectively. The breakthrough times for TCE and benzene were estimated to be 168 days (around five and a half months) and 250 days (around eight and a half months), respectively. The steady-state-permeation rates were estimated to be 0.27 mg/cm^2/day for TCE and 0.14 mg/cm^2/day for benzene. The permeation curves obtained in the aqueous experiments were similar to those obtained in the solvent experiments. Breakthrough of toluene from toluene-saturated aqueous solution did not occur during the ten-month duration of the experiment, after which it was terminated due to a leak in the apparatus. Evidence presented in Chapter 5 predicts that breakthrough would have occurred in approximately 12 months.

To the authors' knowledge, this is the first experimental study to demonstrate that TCE and benzene at concentrations equal to their solubility in water can permeate through PVC pipes from aqueous phase. The breakthrough times in saturated aqueous solutions were much longer than those observed in the neat solvent experiments.

It is known that polymer permeation is directly dependent on the activity of organic compounds in the exposure media. However, the permeation behavior in pure solvents and saturated aqueous solutions showed significant differences in breakthrough times even though the activity for each phase is equal to 1.0 as defined by Berens (1985) and Vonk (1986). Berens (1985) assigned an activity of one to the pure solvent while Vonk (1986) assigned an activity of one to the saturated aqueous solution of the compound. As such, when interpreting the data from various studies, the reference state used in defining the activity must be carefully noted.

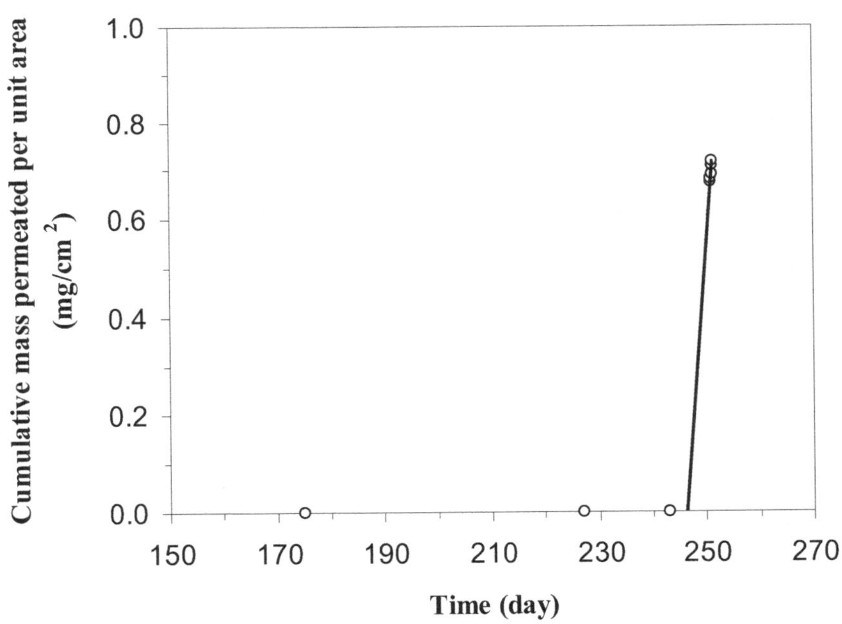

Figure 4.4 Cumulative mass of benzene permeated per unit area for 1-inch PVC pipe exposed to benzene-saturated aqueous solution

To explain the mechanisms involved when a PVC pipe is exposed to a saturated aqueous solution, a key question is whether a constant activity of 1.0 (based on solubility) for organic compounds in aqueous solution is maintained immediately adjacent to the pipe wall. It is possible that mass transfer limitations may occur when organic chemical sorption by the pipe wall is faster than the movement of the organic chemical through the "stagnant water film" surrounding the pipe wall. The permeation process in aqueous solution is not only dependent on the interaction between organic chemicals and polymer materials but also on the availability of compounds for permeation, which may be influenced by the water solubility of chemicals and the diffusion of chemicals in the water. That helps explain why toluene breakthrough lagged behind that of benzene in aqueous solution, while in pure solvents the order of breakthrough times was reversed.

PVC Pipes Exposed to Gasoline

No BTEX compounds were detected in the pipe water after 2 years of exposure. This suggests that new, ungasketed PVC pipes are an effective barrier against commercial gasoline. The resistance of PVC pipes to gasoline permeation was attributed to the relatively low activity of BTEX in gasoline (using the pure solvent as the reference state). Schwarzenbach et al. (1993) defined the activity of compound *i* in a liquid mixture by using fugacity as expressed below:

$$a_i = \frac{f_i}{f_{i,\ pure\ liquid}} \tag{4.1}$$

where a_i = activity of the compound in a liquid

f_i = fugacity of the compound in a liquid mixture

$f_{i,\ pure\ liquid}$ = fugacity of the compound in the pure organic liquid

The fugacity of a chemical in a given phase is equal to the vapor pressure the chemical would have in a gas volume in equilibrium with the phase. The fugacity of a pure organic liquid is equal to its saturated vapor pressure (this is similar to the reference state used by Berens (1985) in defining activity). The vapor pressure of *i* equilibrated with a liquid mixture can be estimated based on Raoult's Law:

$$V_i = \alpha_i x_i P_i^o \tag{4.2}$$

where V_i = vapor pressure of *i* equilibrated with a liquid mixture (atm)

α_i = activity coefficient

x_i = mole fraction of *i* in the mixture

P_i^0 = saturated vapor pressure of *i* in the pure organic liquid (atm)

The activity coefficient α_i is generally set equal to 1.0 if the liquid phase is comprised of a mixture of similar types of organic chemicals and can, therefore, be assumed to be chemically ideal. Under the ideal condition, the activity of compound *i* in a liquid mixture is equal to its mole fraction. Based on this, the activity of BTEX compounds in the commercial gasoline used in this study were estimated according to the gasoline composition and the gasoline molecular weight of 105 g/mole (Reckhorn, 2001). The calculated activities of benzene, toluene, ethylbenzene, m-xylene and o+p-xylene in the gasoline were 0.035, 0.123, 0.016, 0.039 and 0.038, respectively. The total activity of BTEX in the gasoline was 0.251. At this activity, the experimental data showed that permeation of aromatic compounds through PVC pipes was negligible after 2 years, which is consistent with the results of Berens (1985). Berens (1985) indicated that at an activity of 0.25 or less for a single aromatic compound, softening of the PVC will not occur and Fickian diffusion will dominate.

PVC Pipes Exposed to Water Saturated with Gasoline

No BTEX compounds were detected in the pipe water after 2 years of exposure. Since PVC pipe has shown excellent resistance to permeation by free product gasoline, permeation of BTEX from water saturated with gasoline is not expected to occur. Using the measured concentrations of BTEX in gasoline-saturated water and estimating the activity for each BTEX compound (based on the maximal solubility of each compound in water), the total activity of BTEX (sum of individual activity) in gasoline-saturated solution would be 0.248. The experimental data showed no permeation, which corroborates with the findings of Vonk (1986) who suggested that an activity of less than 0.25 (based on maximal solubility) would not result in the permeation of an aromatic compound through PVC.

CONCLUSIONS

Pipe-bottle tests were conducted to investigate the permeation of organic contaminants through PVC pipes under extremely aggressive environments, including neat organic solvents, free product gasoline, saturated organic vapor, and saturated aqueous solutions of solvents and gasoline. Based on the experimental results, the following conclusions can be drawn:

(i) Ungasketed and new PVC pipes were found to be an effective barrier against commercial gasoline. Experimental results also confirmed that PVC was resistant to permeation by BTEX compounds in gasoline-saturated water.

(ii) PVC pipes can be rapidly penetrated when exposed to organic solvents (such as chlorinated solvents and monoaromatic compounds) and their saturated organic vapors.

(iii) Organic contaminants in saturated aqueous solution of organic solvents can also permeate through PVC pipes. This was demonstrated for TCE and benzene and predicted for toluene (Chapter 5). However, the breakthrough times in saturated aqueous solutions were much longer than those observed in pure solvents.

CHAPTER 5
POLYVINYL CHLORIDE (PVC) PIPES – PREDICTIVE TESTS

OVERVIEW

Despite great interest in the susceptibility of PVC pipe to permeation, permeation studies for PVC pipes are few and the quantitative data are extremely sparse. This is usually attributed to the lack of an effective and rapid technique for permeation testing.

The pipe-bottle exposure test presented in Chapters 3 and 4 is one of the conventional techniques for permeation testing. It involves placement of a pipe into the exposure media and periodic analysis of the target compounds in the pipe water. This method has been successfully used for testing PB and PE pipes (Vonk 1985, Park et al. 1991, Holsen et al. 1991b). For PVC pipe, however, use of the pipe bottle exposure test would be a challenge since the permeation of organic molecules through PVC pipe is generally slow, especially for PVC pipe exposed to aqueous solutions (Chapter 4) and at low contaminant levels. To estimate the breakthrough time, long-term sampling and analysis must be carried out, which is time consuming and expensive.

Another technique is the gravimetric sorption test (Berens 1985, Park and Bontoux 1993). This method is based on the equilibrium sorption of a contaminant by a pipe material. The measurement of mass uptake in the sorption test is simple but it yields no information on the breakthrough time of a contaminant. Moreover, the sorption behavior may be strongly dependent on the geometry of polymeric materials and a complicated equation taking into consideration the effect of geometry must be used to analyze the sorption data. To avoid such effect, researchers have used PVC-slab samples (sheet or film) made of pure PVC (Berens 1985). Generally, PVC samples with additives (samples with the same composition as PVC pipe) in a slab shape are unavailable in the market. However, the data obtained from pure PVC-slab samples may not accurately reflect the permeation behavior of chemicals in PVC pipes since the presence of additives in the pipe can modify the state of the polymer at the molecular level and influence the transport properties of the material.

The aim of this study was to apply several predictive techniques to predict the susceptibility of PVC pipe to permeation. The predictive techniques included: (i) gravimetric sorption to investigate the mass uptake of organic molecules by PVC pipe; (ii) microscopic visualization to investigate the formation and propagation of the organic moving front in PVC pipe samples; and (iii) use of near infrared (NIR) reflectance spectroscopy to investigate the change in polymer chain structure in pipe samples due to interactions between pipe materials and organic solvents. Efforts were made to correlate the data obtained from the sorption, moving front (visualization), NIR, and pipe-bottle (Chapter 4) experiments. The implications of these techniques for permeation testing are discussed.

EQUILIBRIUM SORPTION

Materials and Methods

Materials

A total of nine PVC pipes, identified as PVC specimens A to I (PVC SP. A – PCV SP. I), were purchased from local hardware stores. The nine pipes were made by six different manufacturers. The pipe sizes were ¾ inch, 1 inch and 1¼ inch. All pipes were ASTM 1785D-certified and marked NSF-PW.

Reagent grade benzene, toluene, trichloroethylene (TCE) Carbowax® PEG400, 2,2,4,-trimehtylpentane, and n-heptane were purchased from Fisher Scientific (Chicago, IL). Three different types of gasoline, regular gasoline, premium gasoline, and 10% ethanol gasoline (E10) were purchased from local stores.

Experimental Procedures

Sorption experiments were conducted using different PVC pipes under various exposure conditions. The conditions include: neat solvents, mixtures of solvents with polyethylene glycol (PEG), mixtures of solvents with the reference fuel (described below), organic vapors equilibrated with neat solvents, organic aqueous solutions contaminated with solvents, and three types of gasoline (premium, regular and E10). The test matrix is presented in Table 5.1. These exposure conditions generally reflect extreme contaminated field conditions. Sorption under low contamination conditions was not tested since permeation of organic compounds through PVC pipe is negligible for low concentrations of pollutants.

As an inert chemical that does not interact with the PVC pipe material, PEG was used in a previous study (Berens, 1985) to determine the activity threshold at which significant sorption would occur. The National Institute of Standards and Technology's (NIST) formula for "reference fuel" was used in the present study to better simulate the conditions of free product gasoline. The NIST reference fuel consists of 91% (v/v) (91.09% w/w) 2,2,4-trimethylpentane (iso-octane) and 9% (v/v)(8.91 w/w) n-heptane. As a Standard Reference Material (SRM), reference fuel has been used for calibrating instruments and determining lead (NIST 1988) and alcohols (National Bureau of Standards 1986) in gasoline. This reference fuel may be a good surrogate for the alkanes in gasoline which are inert to pipe materials.

Pipe specimens were cut into lengths of 1.5 cm (0.6 inch) with a PVC pipe cutter. The specimens were washed with detergent, rinsed with tap water and distilled water, placed on a paper towel to air-dry, and weighed using an analytical balance.

Solvent Exposure. Specimens were immersed in 50 mL of the test solvent in a 120-mL glass jar with a Teflon-lined lid. For each test condition, the experiment was conducted in triplicate. At various times, the specimens were removed from the solvents, placed on paper towels, wiped, and allowed to air dry for thirty seconds before weighing. Pipe specimens were considered to have reached equilibrium sorption when the range of at least 3 consecutive weighings was no greater than 20 mg (about 0.3 % for a typical 7 g sample).

Vapor Exposure. Preliminary experiments showed that organic vapor condensed on the pipe surface. To avoid the effect of condensation and to simulate the conditions closer to field conditions, a pipe specimen was mounted on the neck of an open glass vial. Approximately 10 mL of deionized water was added to the open glass vial (20 mL) so that the mounted pipe specimen was not in contact with the deionized water. The glass vial was then placed in a 120-mL glass jar filled with 30 mL of the test solvent. Nearly half of the vial was immersed in the solvent, while the other half was above the solvent. The 120-mL glass jar was closed with a Teflon®-lined cap. Under these conditions, the pipe specimens were exposed to organic solvent vapor with humidity close to saturation. For each testing condition, the experiment was conducted in triplicate.

Table 5.1
Test matrix for equilibrium-sorption experiments

Pipe Specimens	Exposure media	Chemicals	Contaminant levels
PVC SP. C/D/I SCH40 (1-inch)	Solvent	Benzene/toluene/TCE	Pure Solvent
PVC SP. D SCH40 (1-inch)	Solvents with PEG	Benzene/toluene/TCE	Pure solvent, 80, 60, 40 and 20% (v/v) in PEG
PVC SP. D SCH40 (1-inch)	Solvents with reference fuel	Benzene/toluene	Pure solvent, 80, 60, 40, and 20% (v/v) in reference fuel
PVC SP. D SCH40 (1-inch)	Vapor	Benzene/toluene/TCE	Saturated organic vapor
PVC SP. D SCH40 (1-inch)	Aqueous	Benzene/toluene/TCE	Aqueous saturated solution (100%), 80%, 60% , 40% , and 20% aqueous solutions of solvents
PVC SP. B/F/H SCH 40 (1.25-inch) PVC SP. C/D SCH 40 (1-inch)	Gasoline	Regular gasoline Premium gasoline E10 gasoline	Free Product

Aqueous Solution Exposure. Three organic compounds (TCE, benzene and toluene) at five concentrations (saturated aqueous solution (100%), 80%, 60%, 40%, and 20% saturated aqueous solutions of the solvents) were tested. Detailed procedures to prepare the aqueous saturated solutions were described in Chapter 4. The saturated aqueous solution was transferred into a 120-mL glass jar and the pipe specimen was then immersed in the solution. This was followed by adding 3-mL solvent into the jar so the aqueous solution was covered with the solvent. The headspace above the solution was minimized as much as possible. Prior to measuring the weight of pipe specimens during the experimental period, the glass jar was opened in a fume hood and the solvent film on the surface was allowed to evaporate. After weighing, the samples were re-immersed in freshly prepared saturated aqueous solution and 3 mL solvent added to the solution. The concentration of the TCE-saturated aqueous solution was maintained at saturation by keeping 3-mL of TCE solvent at the bottom of the jar with regular stirring. In this case, pipe specimens were placed on a glass stand in the jar as to avoid contact with the TCE. For experiments with 80%, 60%, 40% and 20% aqueous solutions of the solvents, the solutions were replaced each week to maintain a relatively constant concentration. A control experiment using deionized water as exposure media was also conducted. For each test condition, the experiment was conducted in duplicate. All experiments were conducted at room temperature (23 ± 1.5°C).

Results and Discussion

PVC Pipes Exposed to Pure Solvents

This study tested three PVC pipes, PVC SP. C, PVC SP. D and PVC SP. I. Results of the equilibrium sorption test for PVC SP. D in pure TCE, benzene and toluene are shown in Figure

5.1. The sorption process was very reproducible and triplicate samples gave identical results when the increased weight was normalized to percent weight gain. Of the three solvents, TCE was absorbed to the greatest extent, followed by toluene and benzene. The approximate time to reach equilibrium for TCE, toluene and benzene was about 4 days, 9 days and 11 days, respectively, and the corresponding percent weight gain (W%) at equilibrium was 70%, 52% and 50%, respectively. The percent weight gains for the three solvents were consistent with those of Berens (1985) but the times needed to reach equilibrium in the Berens study were much shorter because thin PVC specimens were used.

Figure 5.1 also presents the percent weight gain versus the square root of time for the first half of the sorption process. The graph clearly demonstrates that the weight gain was not linear with time but was linear with the square root of time. The results were different from those of Berens (1985) who reported that the percent weight gain was linear with time at high activities (Case II diffusion) and was linear with the square root of time at low activities (Fickian diffusion). A possible explanation for the square-root-of-time behavior observed in this study was the effect of sample thickness. The mechanism involved will be discussed later when the data from the sorption and visualization experiments are compared.

Under the same exposure conditions, all three PVC pipes exhibited the same sorption behavior, although the sorption process for PVC SP. C was found to be slightly faster than that of other two pipes (Figure 5.2). The estimated Flory-Huggins interaction parameters (X_I) of solvent-pipe pairs were almost identical for the three pipe specimens, and were 0.83, 0.81 and 0.86 for benzene, toluene and TCE, respectively. The values of X_I were in the range of 0.5-1.0, indicating that all three organic solvents are strong swelling agents, in which PVC pipe becomes softened or rubbery when exposed to the solvents at room temperature (Behrens 1985).

PVC Pipes Exposed to Solvents with Polyethylene Glycol (PEG)

Three solvents (TCE, benzene and toluene) at five concentration levels (pure solvent, and 80%, 60%, 40% and 20% by volume in PEG were tested in this study.

The percent weight gain versus time for PVC SP. D immersed in pure toluene and its mixtures with PEG are presented in Figure 5.3. Sorption equilibrium was achieved for 100%, 80%, and 60% (v/v) toluene in 9 days, 13 days and 35 days, respectively. Sorption did not reach equilibrium for 40% (v/v) toluene in 2 months of exposure. Extremely slow sorption was found for 20% (v/v) toluene with less than 1% weight gain in 2 months. At the end of the experiments, PVC samples were found to be soft for 40% (v/v) or higher while samples remained rigid at 20% (v/v) toluene.

Figure 5.3(a) presents the percent weight gain versus time and Figure 5.3(b) presents the percent weight gain versus the square root of time for the first 4 days. The plot (Figure 5.3(b)) followed apparent Fickian kinetics (linear with square root of time). The sorption rate ($W\%/day^{1/2}$) is given by the slope of the straight line in Figure 5.3(b). The plot of the sorption rate versus percent (v/v) in PEG (Figure 5.4) demonstrates that the sorption rate decreased dramatically as the toluene concentration (activity) was reduced. When the percent by volume of toluene was below 30%, sorption of toluene from the PEG solutions was insignificant. Using a similar approach, the percent (v/v) benzene and TCE in PEG where sorption was insignificant were also found to be approximately 30%. Sorption of gasoline by PVC pipes would be extremely low since the percent (v/v) monoaromatic compounds is approximately 25% in commercial gasoline. As such, PVC pipes should be an effective barrier against gasoline permeation. This conclusion was supported by the pipe-bottle experiments in Chapter 4.

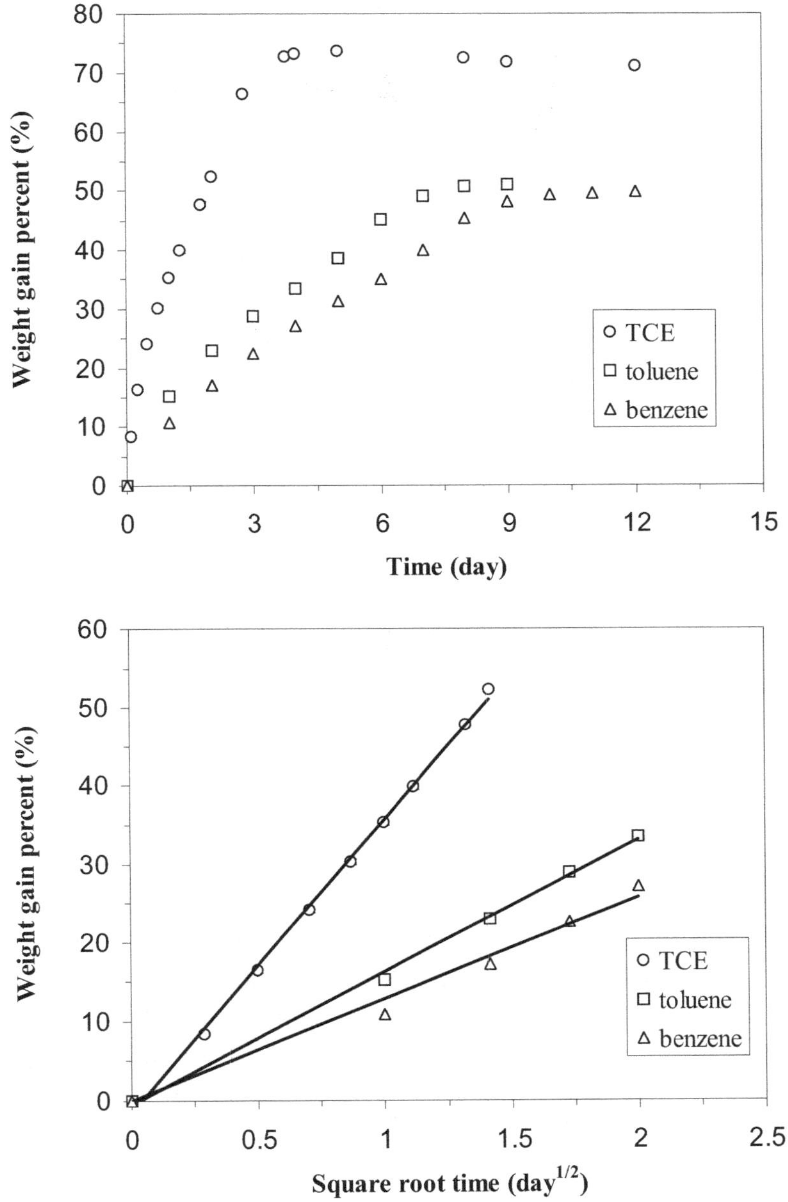

Figure 5.1 Equilibrium sorption tests for PVC SP. D in pure TCE, benzene and toluene

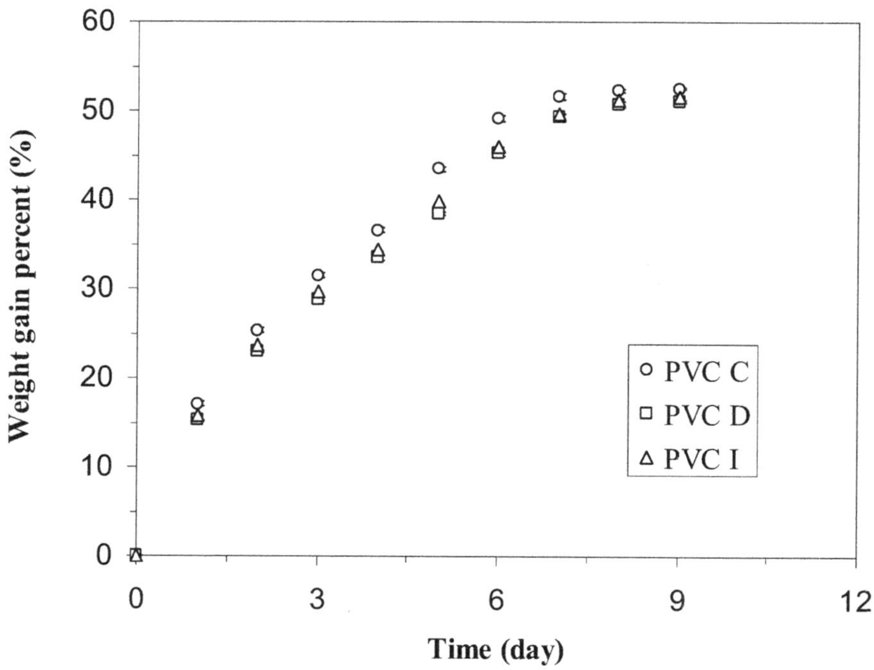

Figure 5.2 Equilibrium sorption tests for PVC SP. C, D and I in pure toluene

PVC Pipes Exposed to Solvents in Reference Fuel

The percent weight gain versus square root of time for PVC D immersed in the mixtures of toluene with the reference fuel is presented in Figure 5.5. Given the identical volumetric fraction of toluene, the sorption rates obtained from the reference fuel experiments were much lower than those from PEG experiments. For example, the sorption rates under the exposure to 80% and 60% (v/v) toluene in reference fuel were found to be 5.3 (%/day$^{1/2}$) and 1.2 (%/day$^{1/2}$), respectively, while the corresponding rates in the mixtures of toluene and PEG were 12.8 (%/day$^{1/2}$) and 9.6 (%/day$^{1/2}$), respectively. At 40% (v/v) toluene, a significant sorption of toluene was observed in the mixture of toluene and PEG, while the sorption in the mixture of toluene and the reference fuel was negligible.

The significant difference in the sorption rates between PEG experiments and the reference fuel experiments may be due to the difference of physicochemical properties of PEG and the reference fuel. Carbowax® PEG 400 is a hydrophilic polymer with an average molecular weight of 380-420 g/mole and a density of 1.1-1.2 g/cm3. In contrast, isooctane, the major component of the reference fuel, is a hydrophobic solvent with a molecular weight of 114 g/mole and a density of 0.69 g/cm3. Theoretically, the sorption process of toluene can be described by the partitioning between the inert chemical (PEG or the reference fuel) and the pipe material. Since the physicochemical properties of toluene are much closer to those of the reference fuel than PEG, toluene would have a greater tendency to partition into reference fuel

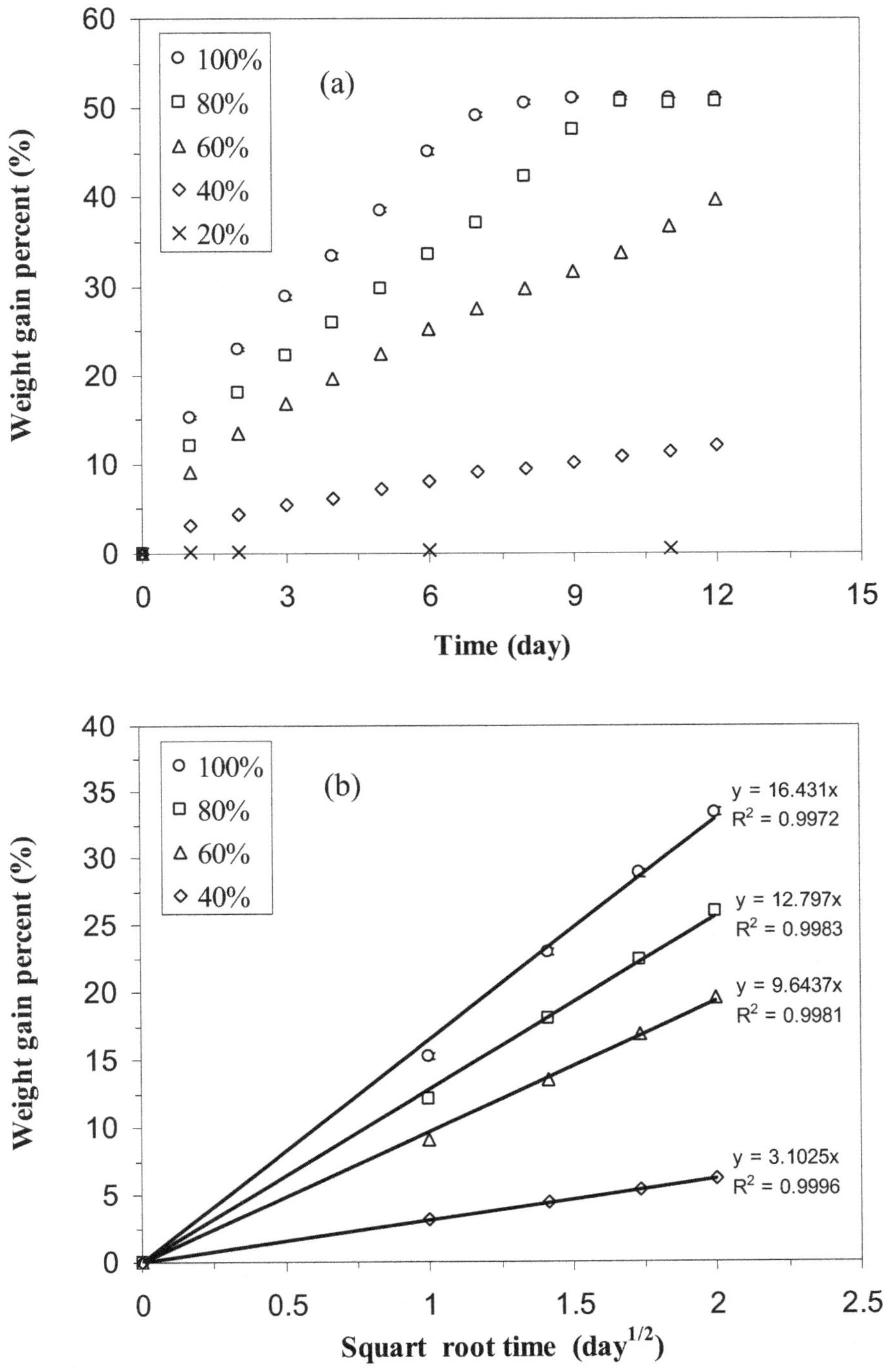

Figure 5.3 Equilibrium sorption tests for PVC SP. D in toluene and polyethylene glycol (PEG)

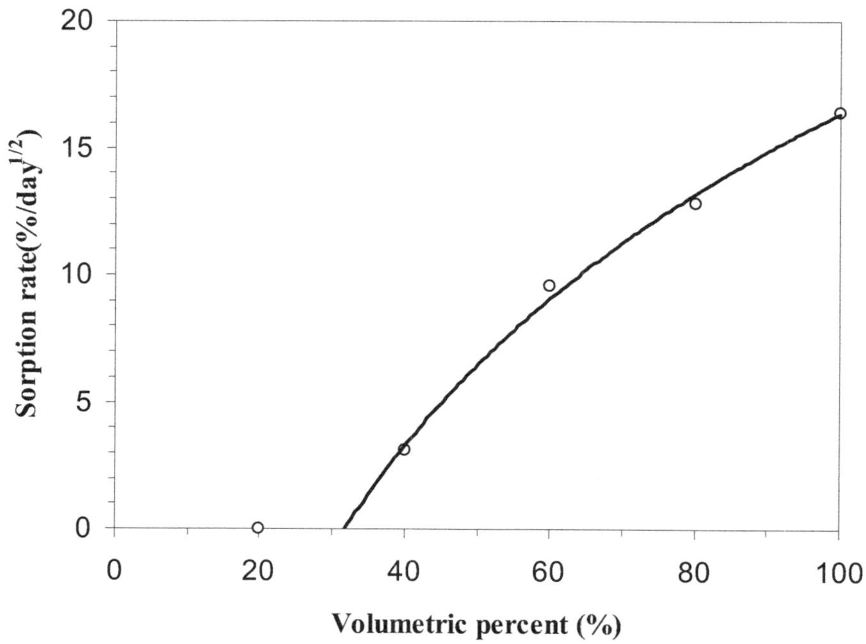

Figure 5.4 Sorption rates vs. percent by volume of toluene in PEG

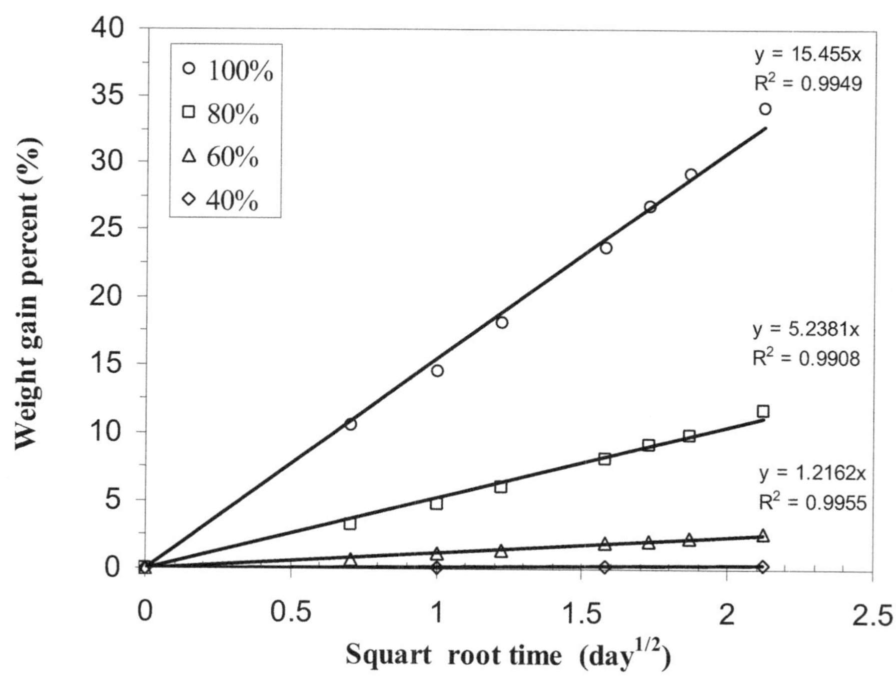

Figure 5.5 Equilibrium sorption tests for PVC SP. D in toluene and reference fuel

rather than into the pipe material. In the mixtures of toluene and PEG, however, toluene may preferentially partition into the pipe material, thereby resulting in greater sorption than from reference fuel. Reference fuel appears to be a better surrogate for gasoline in these experiments because it is more chemically similar to gasoline.

PVC Pipes Exposed to Saturated Organic Vapors

The percent weight gain versus time for PVC SP. D sample exposed to saturated organic vapors of toluene and benzene is presented in Figure 5.6. Similar to the experiments with pure solvents, the data followed apparent Fickian kinetics (linear with square root of time) but the sorption rate was slower. For example, the percent weight gain in saturated benzene vapor (activity of 1.0) was approximately 65% of the weight gain obtained in pure benzene (activity of 1.0) before equilibrium was reached. This is consistent with the data presented by Berens (1985) for toluene at activities of 0.75. The differences found in the present study between the saturated organic vapor and the neat solvent experiments may be due to the effect of mass transfer which will be discussed later when comparing the data obtained from solvent, vapor and aqueous experiments.

PVC Pipes Exposed to Water Saturated with Organic Solvents

The percent weight gain versus time for PVC SP. D immersed in benzene, toluene, and TCE solutions at five saturation levels (100%, 80%, 60%, 40% and 20% saturated aqueous solutions) are presented in Figure 5.7. In comparison with the square-root-of-time relationship in pure solvents or in saturated organic vapor, the percent weight gain was linear with time in aqueous solutions. Sorption in aqueous solutions was much slower than in pure solvents. As shown in Figure 5.7, none of these sorption experiments had reached equilibrium during the 8 months of exposure, although sorption in the 100% benzene-saturated aqueous solution asymptotically approached equilibrium uptake. The sorption rate decreased dramatically as the percent of saturation (contaminant strength) was reduced. For benzene, toluene, and TCE (not shown in Figure 5.7), significant sorption was found for the 100% and 80% aqueous saturation levels while the percent weight gains for 60%, 40%, and 20% saturated aqueous solutions were below 2% after 8 months of exposure. Sorption at the 20% aqueous saturated level was found to be statistically similar to that of the control experiment (deionized water sorption). This implies that sorption can be neglected for contaminant levels below 20% aqueous saturation for both toluene and benzene. Solutions at 20% aqueous saturation (toluene, benzene and TCE at 100 mg/L, 340 mg/L and 220 mg/L respectively) are considered to be a high level of environmental pollution. These concentrations are seldom encountered in the field except in close proximity to a NAPL.

The sorption of an organic compound by polymeric materials is believed to be directly dependent on the activity of the organic compound (rather than the concentration) in the exposure media. Even though the activity of a pure solvent or a compound in saturated aqueous solution or saturated organic vapor is considered to 1.0, the mass sorbed for all three conditions was different. A probable explanation is that mass transfer of the chemicals in water may have an impact. In aqueous solution, sorption of the contaminant was not only dependent on the interaction between organic chemicals and polymer materials but also on the effective availability of the compounds for sorption. It is probable that there is a thin film of stagnant

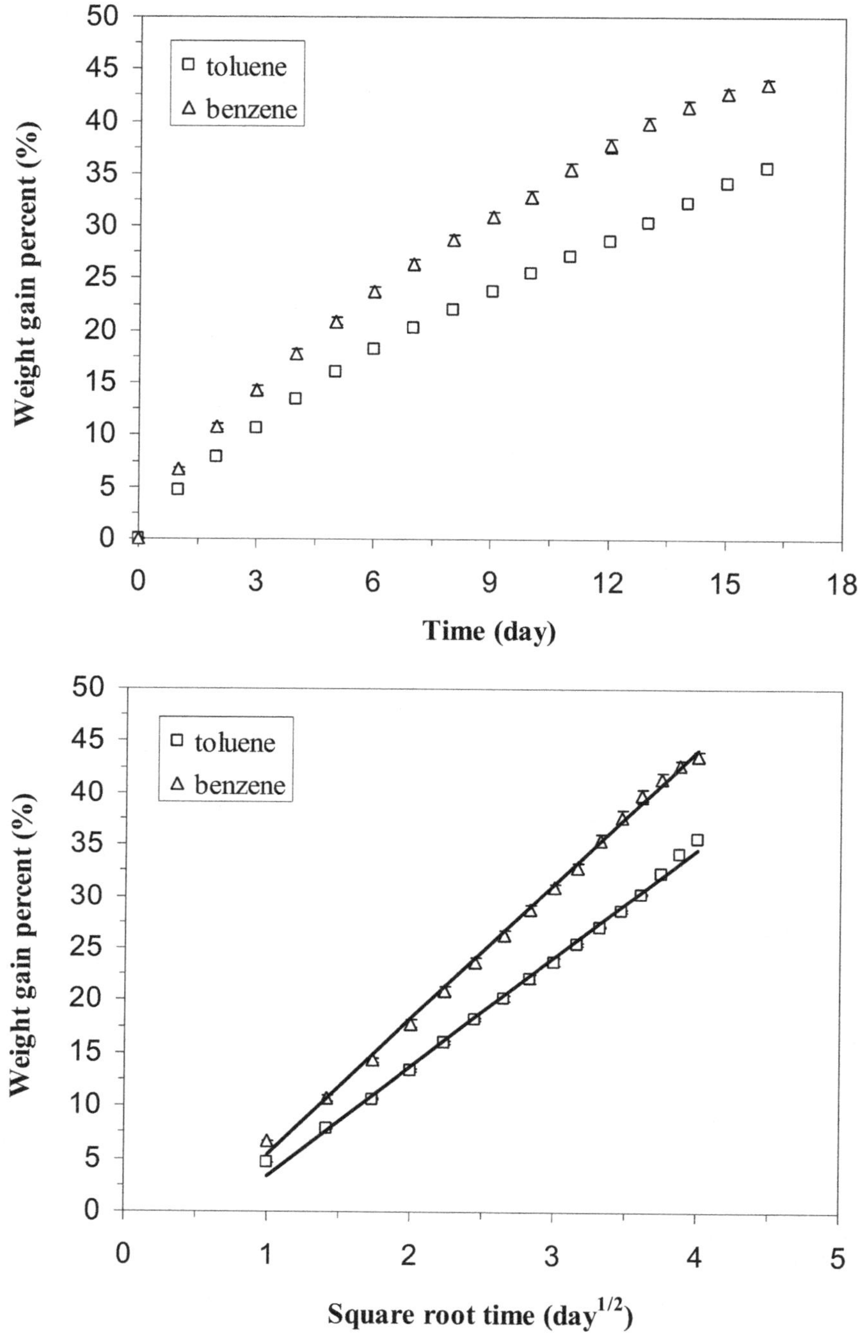

Figure 5.6 Equilibrium sorption tests for PVC SP. D in saturated vapor of benzene and toluene

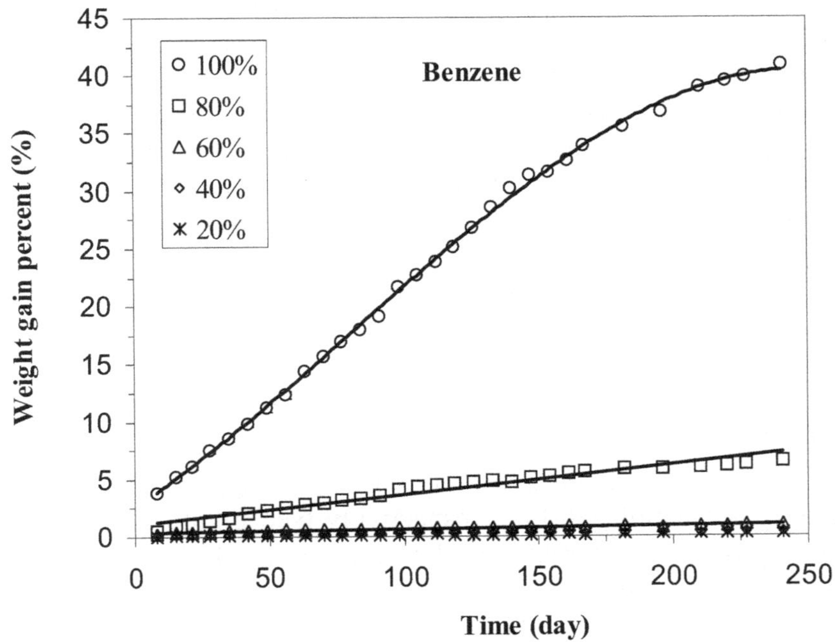

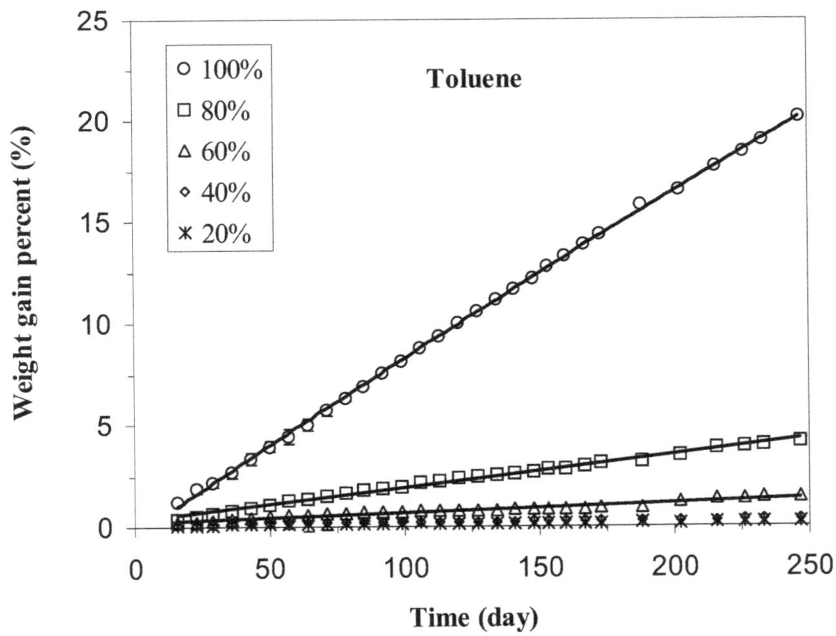

Figure 5.7 Equilibrium sorption tests for PVC SP. D in aqueous solutions of benzene and toluene

water immediately adjacent to the pipe wall for the experiments using saturated aqueous solutions. Organic chemicals must diffuse from the bulk solution through this thin film of stagnant water to the pipe for sorption to occur, even under well mixed conditions. With pure solvent experiments there is direct contact between the pipe and the solvent. In the case of the solvent vapor experiments, the thin film of water may not exist unless the relative humidity is high. Diffusion coefficients through stagnant water are on the order of 10^{-5} cm^2/s, while diffusion coefficients through stagnant air are on the order of 10^{-2} cm^2/s (Schwarzenbach et al. 1993). The difference in the order of magnitudes of diffusion coefficients in air and water may help to explain why sorption in saturated organic vapors was closer to that in pure solvents but much higher than sorption in saturated aqueous solutions, even though the activity of all three solutions were assumed to be equal to 1.0.

PVC Pipes Exposed to Gasoline

After 3 months of exposure to regular, premium, and E10 gasoline, all five PVC pipes (PVC SP. B, C, D, F, H) absorbed the three types of gasoline in very small amounts. The average weight gains at 90 days for all five PVC pipes were 0.30%, 0.73%, and 0.48% for regular, premium, and 10% ethanol gasoline, respectively (see Figure 5.8). The insignificant sorption for gasoline may be due to a relatively low activity of BTEX in gasoline. The total activity of BTEX in the premium gasoline was estimated to be 0.251, with toluene being the highest at 0.123 (Chapter 4). The total BTEX activities for the regular gasoline and the E10 gasoline were 0.128 and 0.119, respectively, with toluene activities of 0.050 and 0.047, respectively. The experimental results are consistent with experiments using the moving front test, described below.

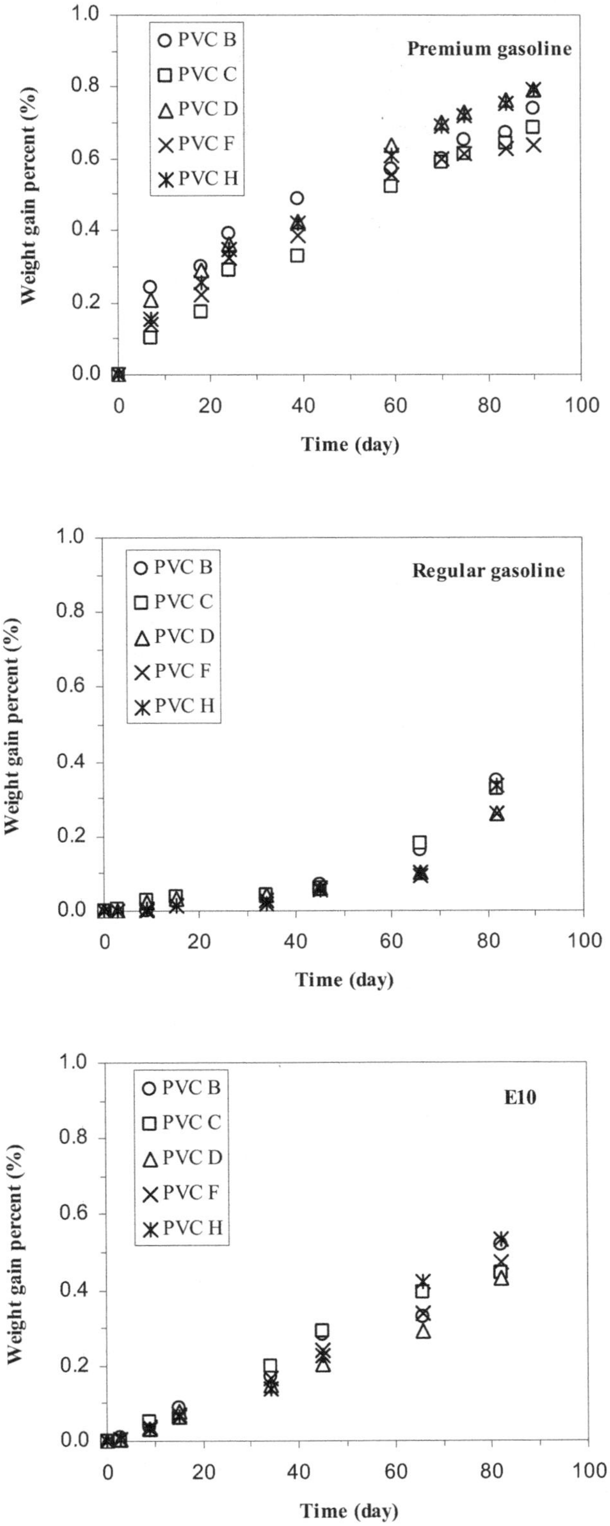

Figure 5.8 Equilibrium sorption tests for pipes in gasoline

MOVING FRONT-VISUALIZATION

Materials and Methods

Materials

Pipes from different manufacturers were purchased from local stores. Reagent grade benzene, toluene, ethylbenzene, xylene (o-xylene, m-xylene and p-xyene), chloroform, carbon tetrachloride, tetrachloroethene (PCE), trichloroethene (TCE), 1,1,2-trichloroethane (1,1,2-TCA), methyl tert-butylether (MTBE), Carbowax® PEG 400, 2,2,4-trimehtylpentane, and n-heptane were purchased from Fisher Scientific (Chicago, IL). Regular gasoline and premium gasoline were purchased from local stores.

Visualization of Moving Front

The approach used here was to observe the formation and propagation of a moving front due to the softening of the PVC by the solvent. Interactions of the swelling solvent with the PVC results in a sharp interface separating the inner glassy core from the outer swollen layer. This interface, defined as the moving front is readily visible under a light microscope, which enables accurate measurement of the advance of the front (Figure 5.9). Three parameters were measured: (i) the original thickness of the pipe (L_0); (ii) the thickness of the swollen layer at time t ($L_{s,t}$); and (iii) the thickness of the remaining glassy core (or the distance from the sharp boundary to the inner wall, $L_{g,t}$). The penetration distance at time t is then calculated by:

$$\text{Penetration distance } (PD_t) = L_0 - L_{g,t} \qquad (5.1)$$

The moving front test includes three procedures (Figure 5.10): (i) sample cutting; (ii) microscopic visualization of the sample cross section with a reflected light microscope (Olympus BHM); and (iii) measurement of the thickness of the swollen layer or the thickness of the remaining glass core using PCI viewer software.

Experimental Procedures

PVC Pipes Exposed to Pure Solvents. Three PVC pipes (PVC SP. C, D, I) and three types of solvents (TCE, benzene and toluene) were tested. The test matrix is presented in Table 5.2. Pipe specimens were cut to approximately 3.5 cm (1.4 ins.) in length. One end of the pipe was sealed with a glass slide using a chemical-resistant epoxy. The pipe was then filled with deionized water and the other end sealed with a glass slide using the same epoxy material. The pipe specimens were then immersed in 50 mL of test solvent in a 120-mL glass jar with a Teflon®-lined lid. At different times, a pipe specimen was removed from the solvent, wiped dry, cut, and the progress of the moving front visualized. The distance of the moving front from the outside to the inside of the pipe (both of $L_{s,t}$ and $L_{g,t}$) was measured.

PVC Pipes Exposed to Solvents with Polyethylene Glycol (PEG). PVC SP. D and three types of solvents (TCE, benzene and toluene) at five concentration levels (pure solvent, and 80%, 60%, 40% and 20% by volume in PEG were tested. PEG is an inert and nonvolatile solvent. The penetration of PEG into PVC is negligible because of its high molecular weight and

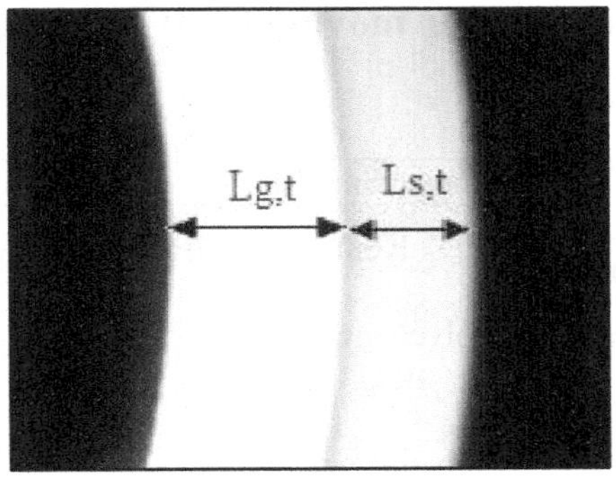

Figure 5.9 Moving front and definition of $L_{g,t}$ and $L_{s,t}$

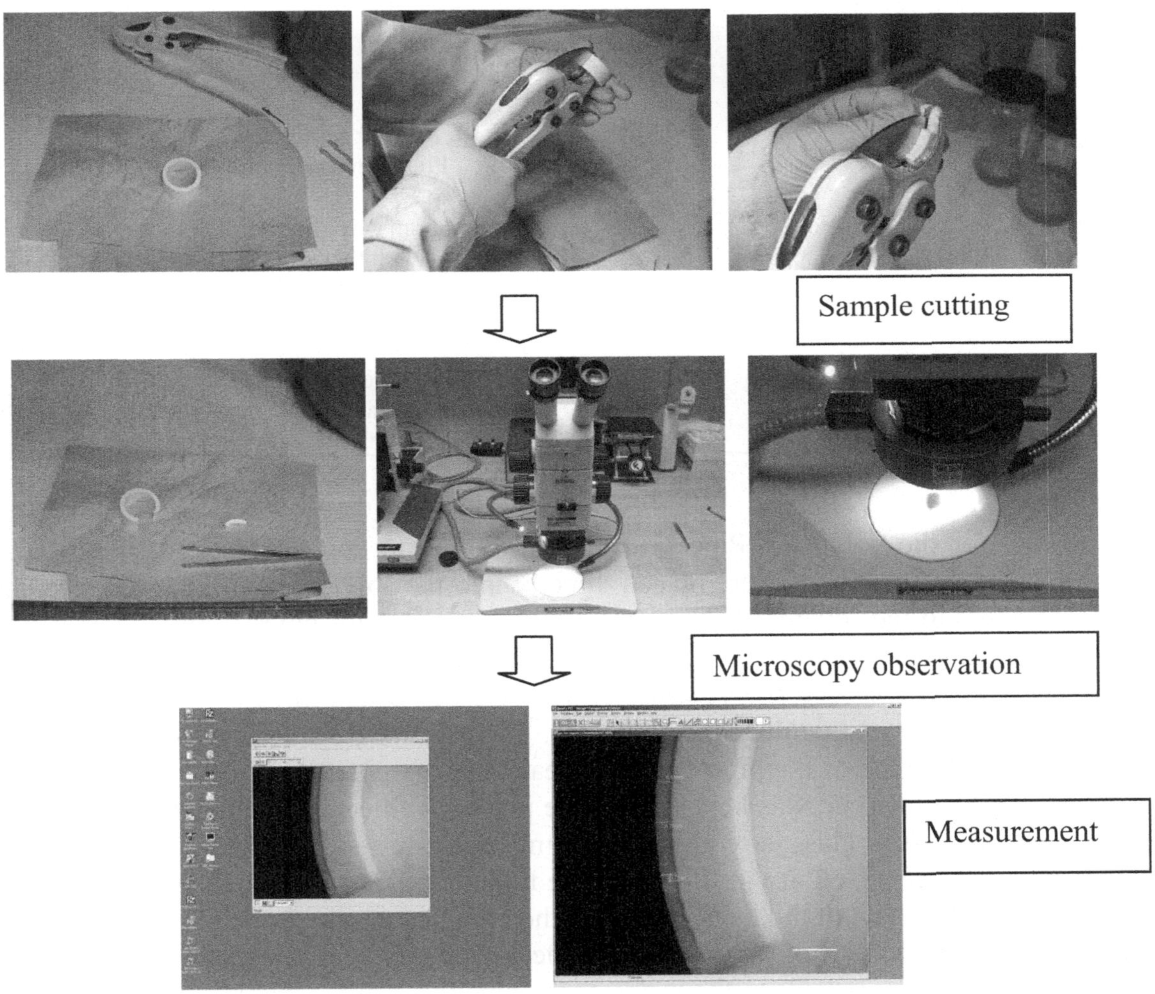

Figure 5.10 Procedures of moving front test

Table 5.2 Test matrix for moving front visualization experiments

Pipe	Exposure media	Chemicals	Contaminant levels
PVC SP. C/D/I SCH40 (1-inch)	Solvent	Benzene/toluene/TCE	-
PVC SP. D SCH40 (1-inch)	Solvents with PEG	Benzene/toluene/TCE	Pure solvent, 80, 60, 40 and 20% by volume in PEG
PVC SP. D SCH 40 (1-inch)	Solvents with reference fuel	Benzene/toluene	Pure solvent, 80, 60, 40, 20% by volume in reference fuel
PVC SP. D SCH40 (1-inch)	Vapor	Benzene/toluene/TCE	Saturated organic vapor
PVC SP. D SCH40 (1-inch)	Aqueous	Benzene/toluene/TCE	Aqueous saturated solution (100%), 80%, 60% , 40% , and 20% aqueous solutions of solvents
PVC SP. C/D/I SCH 40 (1-inch)	Solvent	Regular gasoline Premium gasoline	
PVC SP. D SCH 40 (1-inch)	Solvent	Premium gasoline spiked with toluene	
30 SCH 40 pipes (0.5-2 inch)	Solvent	Toluene	
PVC SP. D SCH 40 (1-inch)	Solvent	4 aromatic compounds 4 halogenated aliphatics MTBE	

low affinity for the polymer. Pipe specimens were cut to approximately 1.5 cm (0.6 ins.) in length. The pipe specimens were placed directly into the solvents with both faces (outside and inside) exposed to the solvent. In the field, a solvent penetrates pipes from the outer wall to the inner wall. Thus, only the outer swollen layer ($L_{s,t}$) was measured.

PVC Pipes Exposed to Organic Vapor. PVC SP. D and three types of saturated organic vapors (TCE, benzene and toluene) were tested. Pipe specimens were cut to approximately 5 cm (2 ins.) in length. One of the pipe ends was sealed with a glass slide using a chemical-resistant epoxy while the other end was kept open. The pipe was filled with deionized water and placed vertically in a 120-mL glass jar containing 30 mL of the test solvent. Approximately 1/3 of the pipe length was immersed in the solvent while the other 2/3 was above the solvent. The water prevented direct contact of the inner wall with the organic vapors. At different times, a specimen was removed and the part of the pipe above the solvent was cut to measure the progress of the moving front from the outside to the inside of the pipe (both of $L_{s,t}$ and $L_{g,t}$).

PVC Pipes Exposed to Aqueous Solutions of Solvents. A series of visualization tests were conducted to investigate the formation and advancement of moving fronts in PVC pipes exposed to aqueous solutions of solvents. PVC D and three organic compounds (TCE, benzene and toluene) at five concentration levels (saturated aqueous solution, 80%, 60%, 40%, and 20%

saturated aqueous solutions of the solvents) were tested. Pipe specimens of about 1.5 cm (0.6 ins.) long were cut and immersed in the prepared aqueous solutions. The procedures to maintain relatively constant contaminant concentrations were described in the previous section on "Equilibrium Sorption." As in the PEG experiments, the pipe ends were not sealed and only the outer swollen layer ($L_{s,t}$) was measured.

PVC Pipes Exposed to Gasoline. Three PVC pipe specimens (PVC SP. C, D, I) were tested. Pipe specimens of 1.5 cm (0.6 ins.) long were cut and then immersed in regular gasoline and premium gasoline. At different times, specimens were removed from the gasoline and the exposed specimens were cut and examined for the presence of a moving front. Another experiment was conducted to determine the BTEX level in gasoline which would result in the formation of moving front after a short-term exposure (one week). In this test, small pipe specimens (1×1 cm (0.4 ins. x 0.4 ins.)) of PVC SP. D were cut and placed into a 40-mL vial. Premium gasoline and pure toluene were added to the vials at the following volume fractions (mL): 40:0, 38:2, 36:4, 34:6, 32:8, 30:10 and 28:12. After one week, the pipe specimens were removed from the solvent and the sections of the exposed specimens were cut and then examined for the presence of a moving front. One vial for each volume fraction was used for gas chromatographic analysis to confirm the concentration of BTEX.

Relative Susceptibility to Permeation of Pipes Obtained from Different Pipe Manufacturers. Thirty PVC pipes were purchased from local stores. The dimensions, specifications and related information (including manufacturer, lot codes, and other information that may be used to trace the specific additives used in compounding of the extruded pipes) were documented. Pipe specimens were cut to approximately 1.5 cm (0.6 ins.) in length, immersed in toluene, and the thickness of the swollen layer ($L_{s,t}$) versus time measured using light microscopy. The propagations of moving fronts for the pipe specimens were statistically compared to assess the relative susceptibility of the different pipe specimens.

Relative Susceptibility to Permeation of PVC-U and PVC-M. The pipes purchased from local stores were PVC-U type which is the dominant PVC material in the US market. Another pipe material, PVC-M, is being used in increasing amounts in Australia and some European countries. PVC-M has a dispersion of cross-linked polyethylene (PEX) throughout the PVC structure and this may impact the resistance to permeation. PVC-M was assessed using the moving front test and the results compared with the relative susceptibility to permeation for PVC-U. A 4-inch diameter PVC-M pipe was obtained from a pipe manufacturer in Australia. Pipe specimens were cut to approximately 1.5 cm (0.6 ins.) in length and immersed in four solvents (toluene, benzene, TCE, chloroform), and the corresponding swollen layer thickness ($L_{s,t}$) versus time was measured using light microscopy. The same tests were conducted for a PVC-U pipe and the results compared with that of PVC-M.

Prediction of the Behavior of Organic Solvents Penetrating through PVC Pipes Based on Some Physiochemical Properties of the Solvent. Nine organic solvents were selected on the basis of their occurrence in contaminated soil and groundwater. The solvents can be classified into three groups: aromatic hydrocarbons, halogenated aliphatics, and oxygenated hydrocarbons. Pipe specimens were cut to approximately 1.5 cm (0.6 ins.) in length and immersed in the 9 types of solvents. The swollen layer thicknesses ($L_{s,t}$) versus time were measured using light microscopy. Efforts were made to correlate the penetration rates with various properties of organic compounds such as the octanol-water partition coefficient, molecular diameter, and the Flory-Huggins interaction parameter. All experiments above were conducted in duplicate or in triplicate at room temperature (23 ± 1.5°C).

Results and Discussion

PVC Pipes Exposed to Pure Solvents

In pure solvents, a sharp boundary (moving front) separating the inner glassy core from the outer swollen layer is distinctly observable as the solvent advances into the PVC matrix. This offers an opportunity to carry out a dynamic observation of the progression of the moving front with time. Figure 5.11 shows the advance of the moving front in PVC SP. D exposed to pure toluene. The moving front reached the inner wall on the 16th day, the day when the breakthrough of toluene was detected in the pipe water (Figure 4.2 in Chapter 4). The results indicated that no detectable permeation occurred until the moving front reached the inner wall of the PVC pipe. To the authors' knowledge, this is the first experimental study directly demonstrating the relationship between the moving front and permeation in PVC pipe.

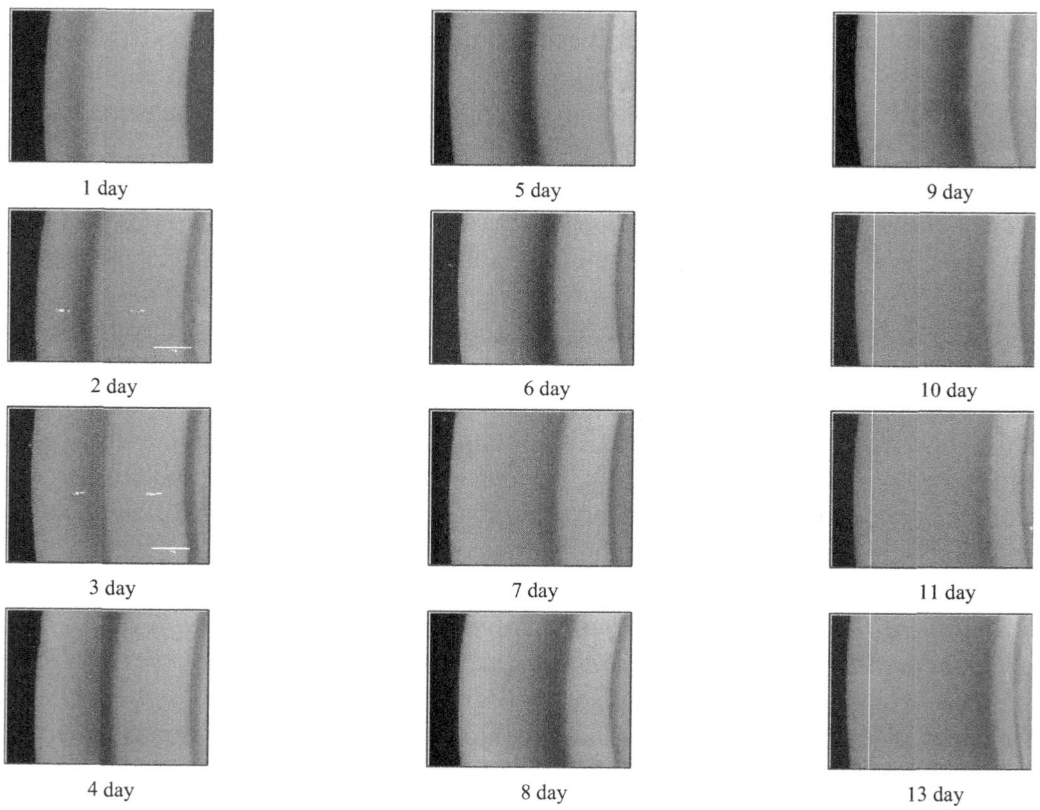

Figure 5.11 Advance of moving front with time in PVC SP. D exposed to pure toluene (left line: boundary of swollen outer wall; mid line: moving front; right line: boundary of inner wall)

Figure 5.12 presents the results of the moving front tests for three pipes (PVC SP. C, PVC D and PVC I) in pure toluene. All three PVC pipes exhibited the same penetration behavior with the solvent penetrating the PVC SP. C slightly faster than the other two pipes. As shown in the previous sorption test (Figure 5.2), PVC SP. C also exhibited a slightly more rapid sorption behavior. These results suggest that the sorption process and the penetration (moving front) process must be strongly correlated. Such correlation was further demonstrated by the later experiments using toluene in PEG.

Results of the moving front test for PVC SP. D in pure TCE, benzene and toluene are presented in Figure 5.13. The progress of the moving front with time was very reproducible with triplicate samples giving identical results. TCE penetrated PVC pipe the most rapidly, followed by toluene and benzene. The thickness of the swollen layer and the penetration distance were found to be linear with the square root of time. The penetration distance as a function of the square root of time can be used to predict the breakthrough time of an organic solvent through the PVC pipe.

Using the regression equations listed in Figure 5.13, the breakthrough times (T) for a pipe wall thickness of 3.61±0.02 mm (1 in. pipe) for the three solvents were predicted to be:

$$T = \left(\frac{3.61 + 0.3695}{0.3109}\right)^2 = 164\ hours = 6.8\ days \text{ for TCE} \tag{5.2}$$

$$T = \left(\frac{3.61 + 0.3451}{1.0309}\right)^2 = 14.7\ days \text{ for toluene} \tag{5.3}$$

$$T = \left(\frac{3.61 + 0.2148}{0.8789}\right)^2 = 18.9\ days \text{ for benzene} \tag{5.4}$$

The pipe-bottle test indicated that the actual breakthrough occurred at about 6.5 days for TCE, 16 days for toluene, and 20 days for benzene (see Figure 4.2 in Chapter 4). The results revealed that the breakthrough time can be predicted by using the penetration data obtained from the moving front test, which can be considered a predictive test.

As shown in Figure 5.13, the penetration of the front was characterized by square-root-of-time dependence rather than linear time dependence. This observation deviated from a key feature of Case II diffusion, i.e., constant velocity of the moving front (Thomas and Windle 1982; Hui et al. 1987a, 1987b). The previous sorption curve (Figure 5.2) also showed a linear increase of polymer weight gain as a function of the square root of time rather than a linear time dependence. Before discussing the mechanisms involved, a fundamental issue needs to be examined: what are the "criteria" for Case II diffusion for a geometry of a cylinder? Specifically, would a penetrating solvent front maintain a constant rate throughout the cylindrical polymer sample for conditions where the solvent penetrating behavior in a flat sheet sample is Case II diffusion? Lee and Kim (1992) investigated the effect of geometry on solvent penetration in glassy polymers, and they found that solvent penetration in radially symmetric geometries exhibited the same transport characteristics as in sheet samples, except that at later stages, penetration of the solvent was observed to be accelerated towards the core. Therefore, if solvent penetration in PVC pipe conforms to Case II diffusion, the moving front would advance

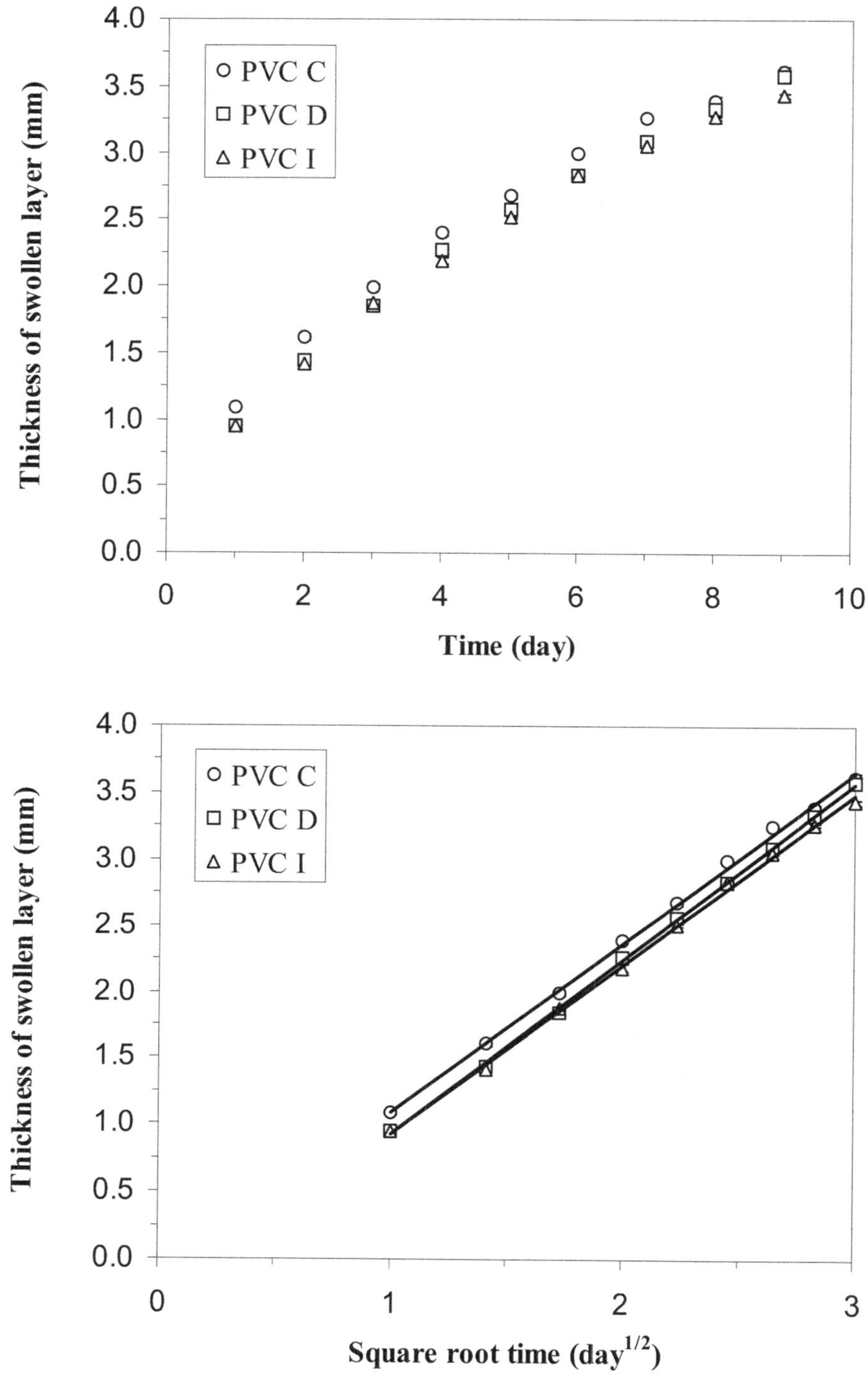

Figure 5.12 Moving front tests for PVC SP. C, D and I in pure toluene

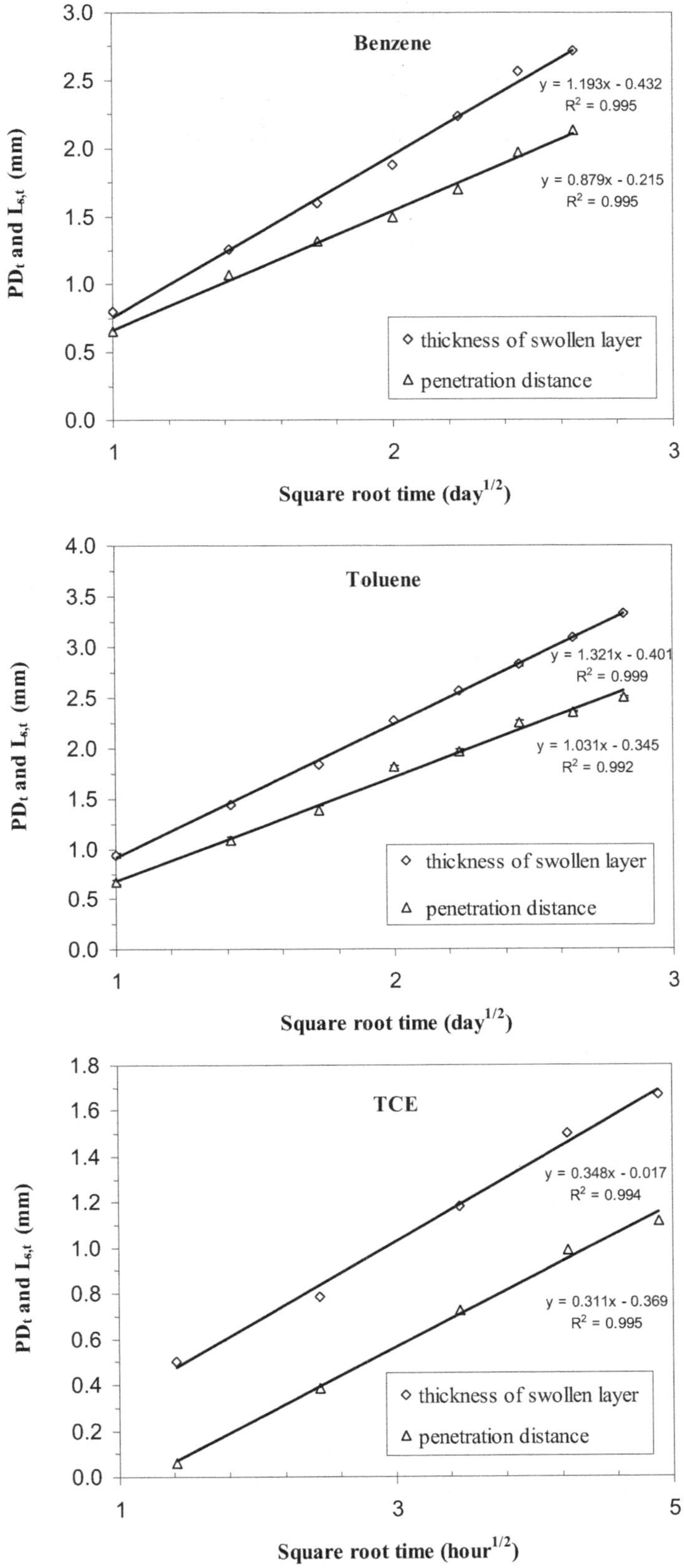

Figure 5.13 Moving front tests for PVC SP. D in pure TCE, benzene and toluene

at a constant velocity, at least in the initial stage. The results from this study refute the proposed Case II diffusion mechanism for the solvent-penetrating behavior in PVC pipes (Berens 1985, Vonk 1986).

A possible explanation for the square-root-of-time behavior observed in the experiments is the effect of sample thickness. Since Case II diffusion represents a shift in balance between the solvent diffusion and polymer relaxation processes with the latter being rate-controlling, increasing the sample thickness inevitably will result in a diffusion-limited situation, thereby shifting the linear kinetics to the familiar square-root-of-time Fickian behavior (Hopfenberg 1978, Astarita and Joshi 1978). Generally, the wall thickness of commercial PVC pipes (> 3 mm) is much greater than those of thin sheet specimens (< 1 mm) used in the studies by others where Case II diffusion was found.

PVC Pipes Exposed to Solvents with PEG

The thicknesses of the swollen layer ($L_{s,t}$) versus the square root of time for PVC samples (PVC SP. D) exposed to toluene at five concentration levels (pure solvent, 80%, 60%, 40% and 20% (v/v) in PEG) are plotted in Figure 5.14. Consistent with the findings for pure solvents, the thickness of the swollen layer was linear with the square root of time in the mixtures of toluene and PEG. The growth rates of the thickness of swollen layer decreased sharply with decreasing toluene concentration (Figure 5.15). When the percent by volume was below 25%, the rate was nearly zero, which means no moving front will be formed for this contaminant level or lower. This is supported by observations that no moving front was detected in PVC pipe samples exposed to 20% (v/v) toluene in PEG for 8 months. For commercial gasoline, the total volumetric percentage of aromatic compounds (BTEX) was approximately 25% with toluene being the highest at 12.3%. Therefore, the moving front results here indicated that the penetration of gasoline in PVC pipes would be extremely slow and PVC pipe should be an effective barrier against gasoline permeation. This is consistent with the results of the pipe-bottle experiments in Chapter 4, where no BTEX was detected in PVC pipe exposed to gasoline for nearly two years.

Figure 5.16 shows the thickness of the swollen layer versus the corresponding weight gain obtained from the toluene and toluene-PEG-mixture experiments. It is not surprising that there was a strong linear relationship (r^2 >0.99) between the weight gain and the thickness of swollen layer. Sorption leads to an increase in the solvent concentration at the polymer surface and, thus to the formation of a moving front. An increase in the solvent sorbed would result in an increase in swelling and would promote advancement of the moving front.

PVC Pipes Exposed to Solvents with the Reference Fuel

The thicknesses of the swollen layer (Ls,t) versus the square root of time for PVC samples exposed to toluene at four concentration levels (pure solvent, 80, 60, and 40% by volume in the reference fuel) are plotted in Figure 5.17. In comparison with the results obtained from the PEG experiments, toluene penetrated into pipe samples at much slower rates in the mixtures of toluene and the reference fuel. For example, the growth rate of the thickness of swollen layer was found to be 0.445 (mm/day$^{1/2}$) and 0.127 (mm/day$^{1/2}$) under the exposure of

80% and 60% by volume of toluene in the reference fuel experiments respectively, while the corresponding rate

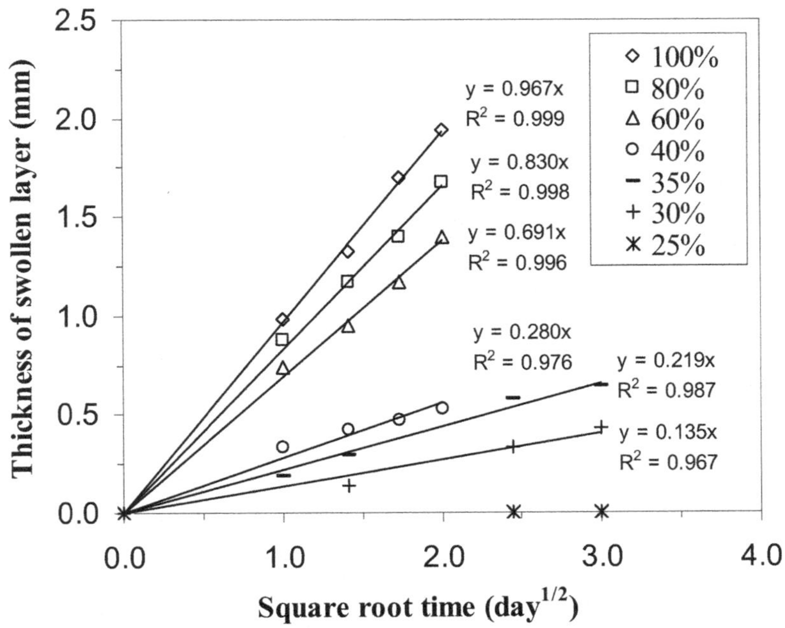

Figure 5.14 Moving front tests for PVC SP. D by % toluene (v/v) in PEG

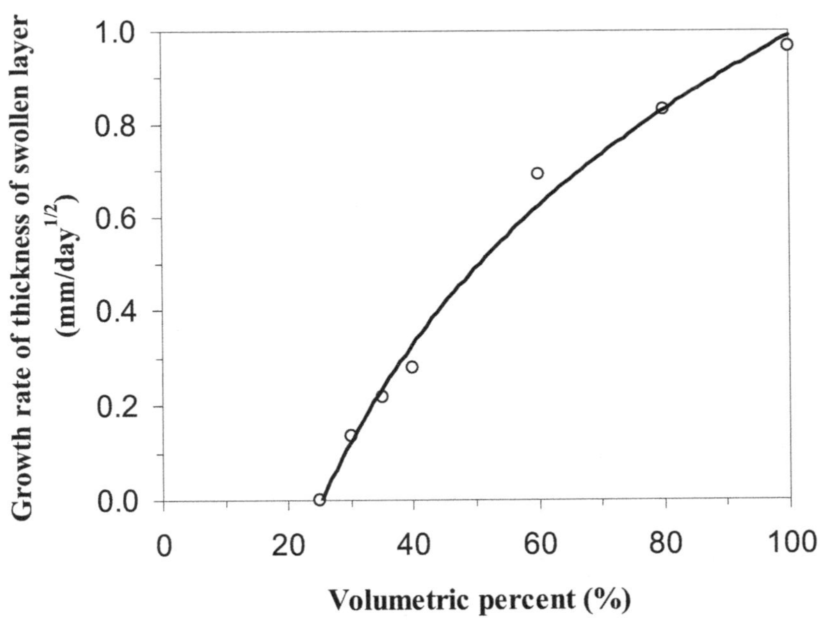

Figure 5.15 Growth rates of swollen layer vs. % (v/v) toluene in PEG

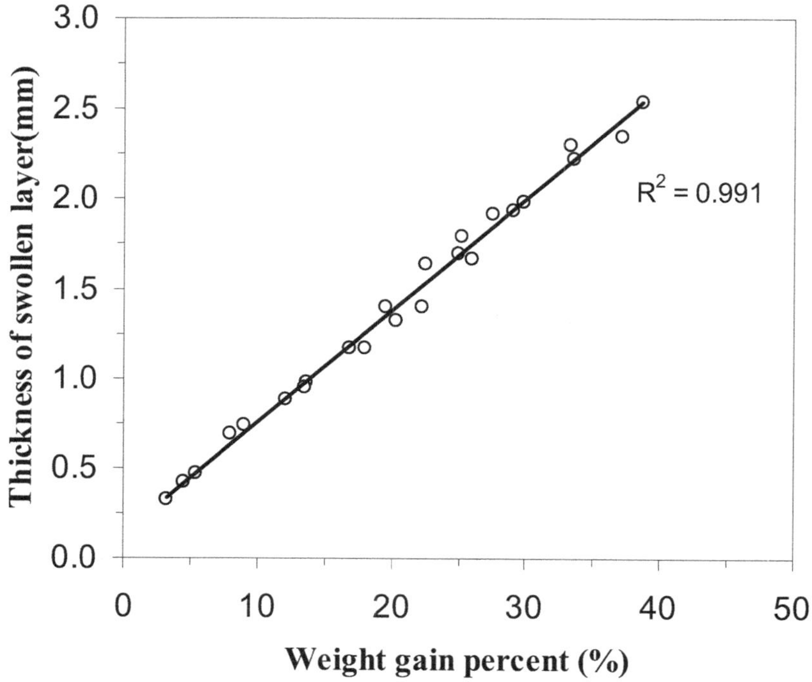

Figure 5.16 Correlation of weight gains with the thickness of swollen layer for PVC SP. D immersed in pure toluene and its mixtures with PEG

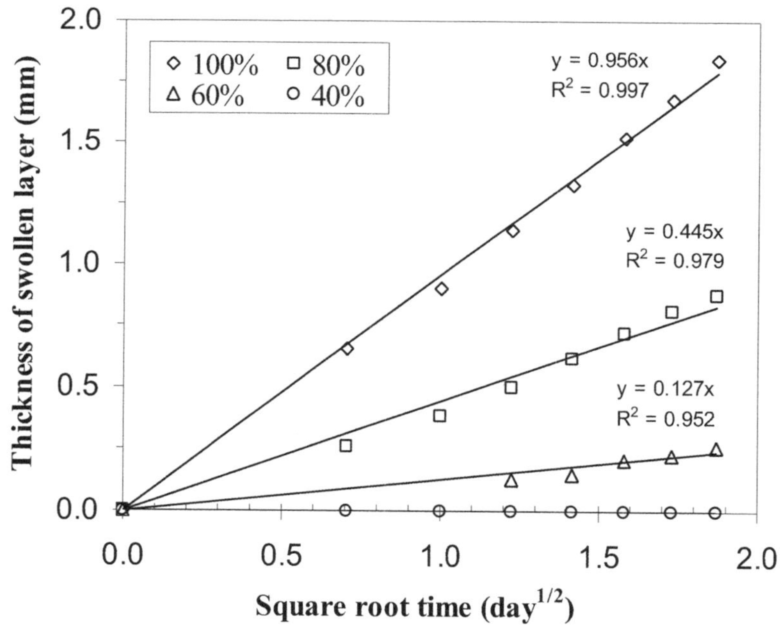

Figure 5.17 Moving front tests for PVC SP. D in toluene with the reference fuel

was 0.830 (mm/day$^{1/2}$) and 0.691 (mm/day$^{1/2}$) in the mixtures of toluene and PEG respectively. At 40% (v/v) toluene, a significant penetration of toluene was observed in the mixture of toluene and PEG while no moving front was detected in the mixture of toluene and the reference fuel. The significant difference in the moving front data between PEG experiments and the reference fuel experiments may be due to the difference of physicochemical properties between PEG and reference fuel. This effect was been discussed above when comparing the sorption data obtained from the reference fuel experiments and the PEG experiments.

Compared to the reference fuel experiments, PEG experiments provided a conservative estimation of the strength threshold at which permeation would not occur in PVC pipes. The 25 percent by volume of toluene was found as the threshold value in PEG experiments while the reference fuel experiments revealed that the threshold value was as high as 40% (v/v) toluene (Figure 5.18). To evaluate the susceptibility of PVC pipe to permeation in free product gasoline, however, there is a remaining question: will the penetration characteristics of BTEX mixtures differ from those obtained from a single compound (such as toluene) studies? More specifically, can the threshold value for a single compound be applied to mixtures of compounds?

Figure 5.19 presents the thicknesses of the swollen layer ($L_{s,t}$) versus the square root of time for PVC samples exposed to benzene, toluene, ethylbenzene, xylene, and the mixture of BTEX with 25 percent by volume for each compound. As reported previously, no moving front was detected for a single compound with 25% by volume. However, the mixture of BTEX was sufficiently aggressive to produce the moving front and penetrate into the pipe matrix. The mixture of BTEX also displayed a "compromised" aggressiveness since it was more aggressive than ethylbenene and xylene but less aggressive than benzene and toluene. Moreover, the moving front data for BTEX at a specific exposure time can be simply modeled by multiplying the factor of 0.25 to the measured moving front data for each single compound and then adding all the obtained values together. It demonstrated that the impacts of organic mixture on PVC pipes are additive.

Further work was conducted to investigate the formation and propagation of the moving fronts in PVC samples exposed to the BTEX mixtures at five concentration levels in the reference fuel (total 100%-25% for each compound by volume, total 80%-20% for each compound by volume, total 60%-15% for each compound by volume, total 40%-10% for each compound by volume and total 20%-5% for each compound by volume). Similar to the observations in the mixtures of toluene and the reference fuel, no moving front was detected when the volumetric percent of the total BTEX was 40% (10% for each compound) or lower (Figure 5.20).

PVC Pipes Exposed to Organic Vapor

Results of the moving front test for PVC SP. D in saturated benzene vapor are presented in Figure 5.21. Similar to the observations for pure solvents, the data followed apparent Fickian kinetics (linear with square root of time) but the growth rate of the thickness of swollen layer was relatively slow. The penetration distance in saturated benzene vapor was approximately 60% of the distance obtained in pure benzene for a specific exposure time. In our previous sorption experiments, the weight gain for the saturated benzene vapor was approximately 65% of the weight gain for exposure to pure benzene. This shows that the moving front and the sorption experiments gave similar results. The difference between the saturated organic vapor and the neat solvent is probably due to the mass transfer effect. This effect has been discussed

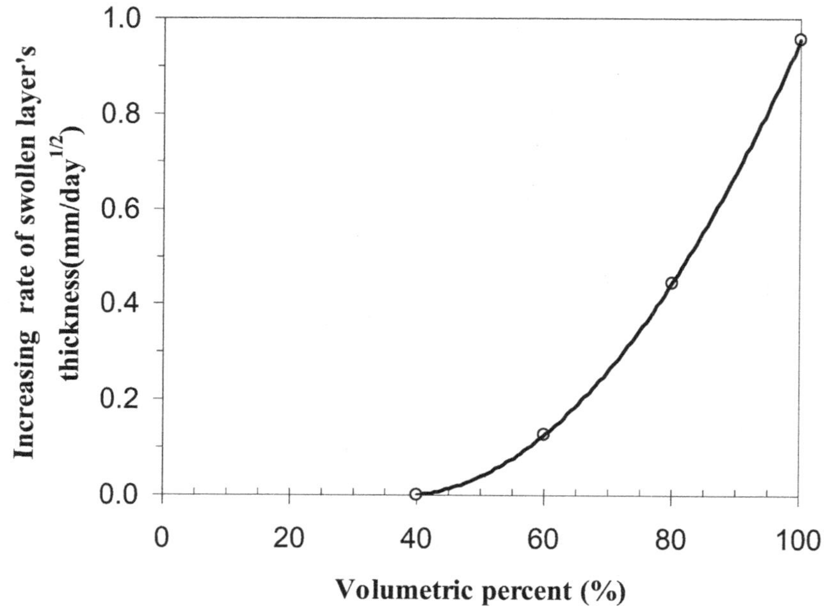

Figure 5.18 Growth rates of the thickness of swollen layer vs. percent by volume of toluene (data from the reference fuel experiments)

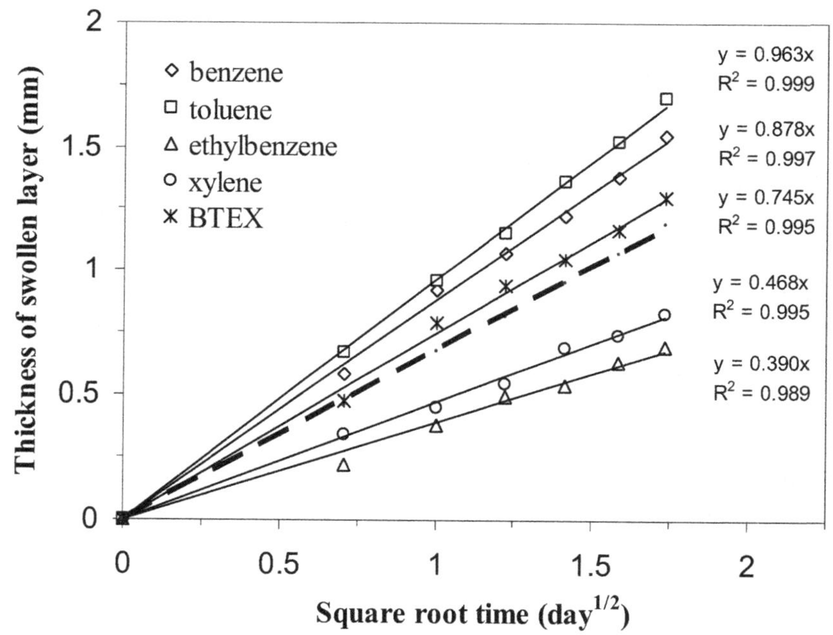

BTEX: mixture of 25 percent by volume of benzene, toluene, ethylbenzene and xylene
Data for dashed line were calculated by: $0.25 \times M_{b,t}+0.25\times M_{t,t}+0.25\times M_{e,t}+0.25\times M_{x,t}$;
$M_{b,t}$, $M_{t,t}$, $M_{e,t}$ and $M_{x,t}$ were the moving front data at a specific time t for benzene, toluene, ethylbenzene and xylene, respectively.

Figure 5.19 Moving front tests for PVC SP. D in benzene, toluene, ethylbenzene, xylene, and their mixture

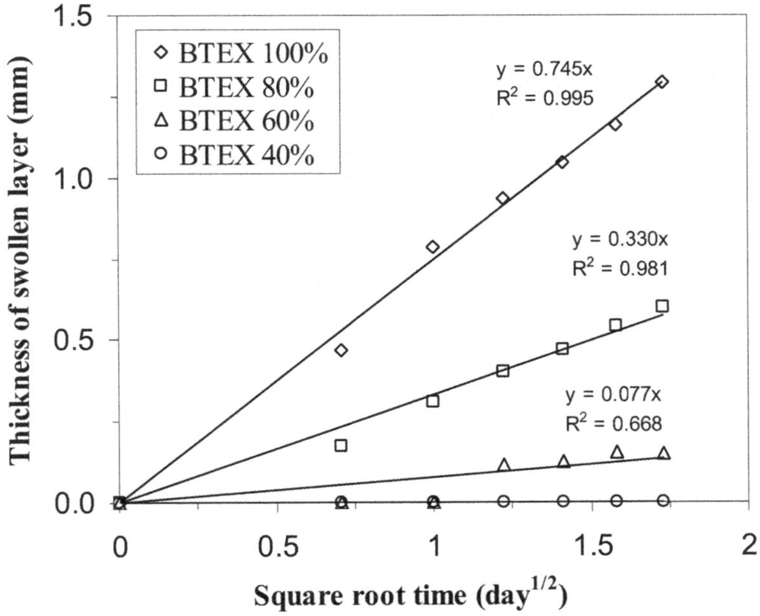

BTEX 100%: 25 percent by volume of each BTEX compound
BTEX 80%: 20 percent by volume of each BTEX compound
BTEX 60%: 15 percent by volume of each BTEX compound
BTEX 40%: 10 percent by volume of each BTEX compound

Figure 5.20 Moving front tests for PVC SP. D in BTEX mixtures with reference fuel

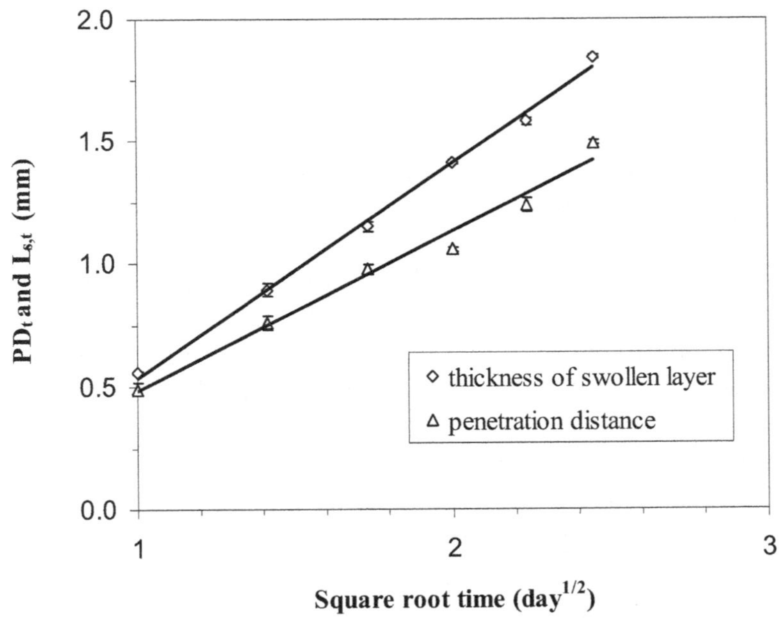

Figure 5.21 Moving front tests for PVC SP. D in saturated benzene vapor

previously when comparing the sorption data obtained from solvent, vapor, and aqueous tests.

PVC Pipes Exposed to Organic Aqueous Solutions

The formation of a moving front was found to be strongly dependent on the weight gain from aqueous solution. In this study, the critical weight gain which would result in the formation of an observable moving front was estimated to be roughly 2-3%. In the 100% saturated aqueous solutions of benzene and TCE, a moving front appeared during the first two weeks of exposure. In contrast, there was an induction period for the formation of the moving front in the 100% saturated aqueous solution of toluene, mainly due to relatively low uptake of toluene. The moving front was also detected in the 80% saturated aqueous solutions once the weight gain exceeded 2%. No moving front was found in the other three solutions (60%, 40% and 20% saturated aqueous solutions) after 12 months of exposure.

The thickness of the swollen layer versus time for aqueous saturated solutions of toluene and benzene are shown in Figure 5.22. Generally, the thickness of the swollen layer increased linearly with time during the initial period. A delay in the swelling was observed for PVC pipes in the solutions of toluene. This delay in the formation of the moving front was probably due to the relatively lower solubility of toluene in water as compared to benzene. In the experiments with pure solvents, the thickness of a completely swollen 1-inch PVC sample was found to be about 4.8 mm. If the same thickness for a completely swollen sample in aqueous saturated solutions was assumed, the breakthrough time in the saturated solution can be estimated based on the rate of swelling obtained from Figure 5.22. Based on this reasoning, the predicted breakthrough times for an aqueous saturated solution of benzene and toluene were 8 months and 12 months, respectively. The pipe-bottle test indicated that the actual breakthrough in an aqueous saturated solution of benzene occurred approximately 250 days (about 8.3 months) (Figure 4.5 in Chapter 4) and the breakthrough of toluene did not occur within ten months, after which the experiment was terminated due to a leak. The results revealed that the breakthrough time for an aqueous solvent-saturated solution can be predicted by using the penetration data obtained from this predictive test.

PVC Pipes Exposed to Gasoline

No moving front was detected for the pipes exposed to premium gasoline and regular gasoline within a one-year test period. The results further confirmed that PVC pipe is an effective barrier against gasoline permeation.

Generally, BTEX in gasoline is roughly 25% or lower by volume. However, the fraction of BTEX in gasoline depends on the source of the crude oil and the refining process. For example, the fraction of monoaromatics (BTEX) in catalytically refined gasoline can be more than 50% (Rittmann and McCarty, 2001). To comprehensively evaluate the potential permeation risk for gasoline, the critical BTEX level in gasoline, which would result in the formation of moving front, should be known. To determine this level, PVC SP. D samples were exposed to premium gasoline with varying toluene concentrations by adding pure toluene to the gasoline. The compositions of the mixtures are presented in Table 5.3. The results of moving front test after one week of exposure are shown in Figure 5.23.

The moving front was only detected for total BTEX $\geq$ 0.43% (v/v) and with toluene $\geq$ 31% (v/v). No moving front was detected for total BTEX $\leq$ 32% (v/v) and toluene $\leq$ 21% (v/v)

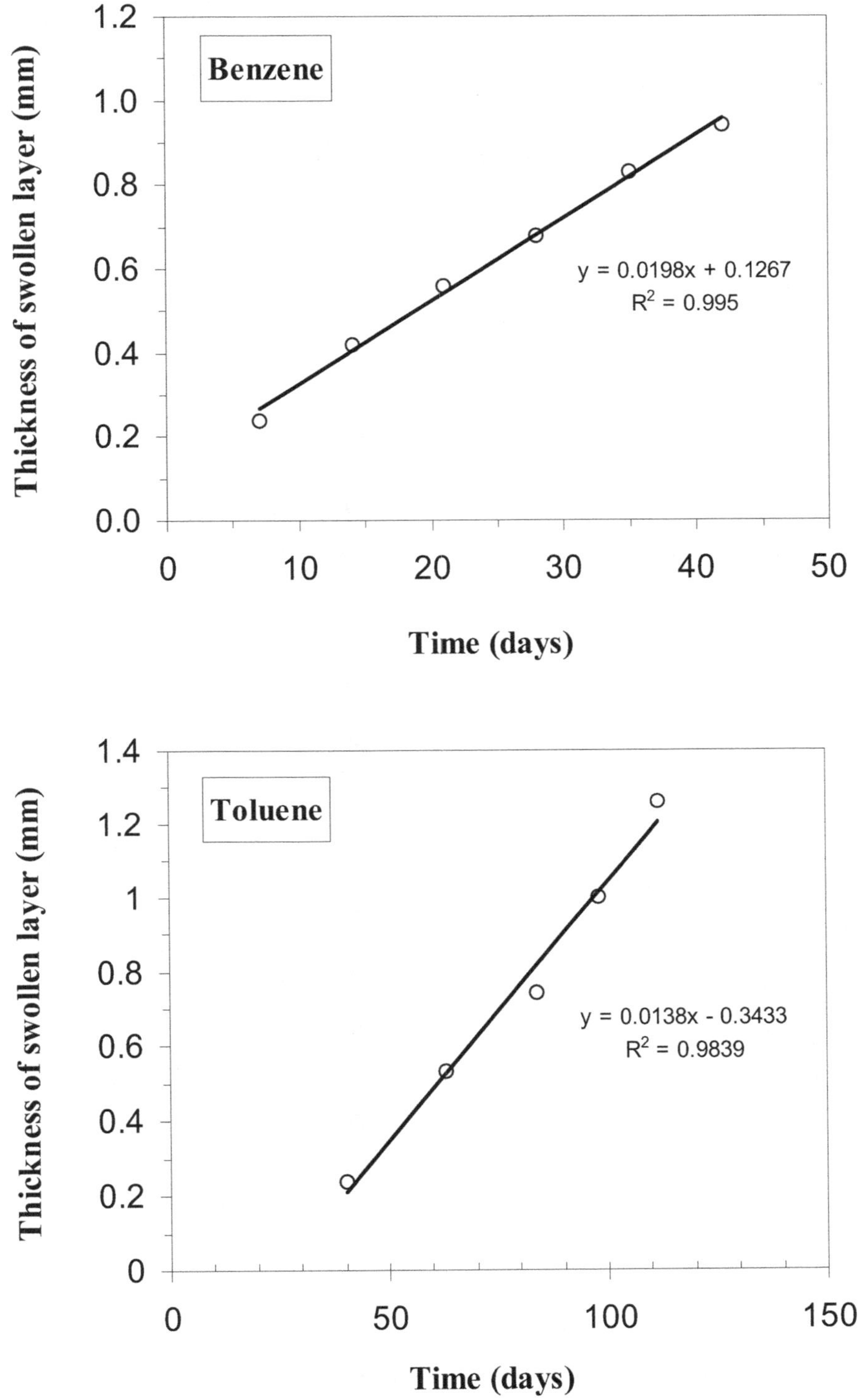

Figure 5.22 Moving front tests for PVC SP. D in saturated aqueous solutions of benzene and toluene

Table 5.3
Compositions of mixtures of premium gasoline and toluene

	A	B	C	D	E	F	G
	Volumetric fraction in gasoline						
Benzene	0.034	0.030	0.025	0.027	0.023	0.020	0.018
Toluene	0.135	0.183	0.212	0.308	0.327	0359	0.377
Ethylbenzene	0.018	0.015	0.013	0.015	0.012	0.011	0.010
m-Xylene	0.051	0.043	0.036	0.041	0.033	0.029	0.024
O+p-Xylene	0.050	0.04	0.036	0.040	0.032	0.029	0.023
Total BTEX	0.287	0.314	0.322	0.431	0.427	0.448	0.451

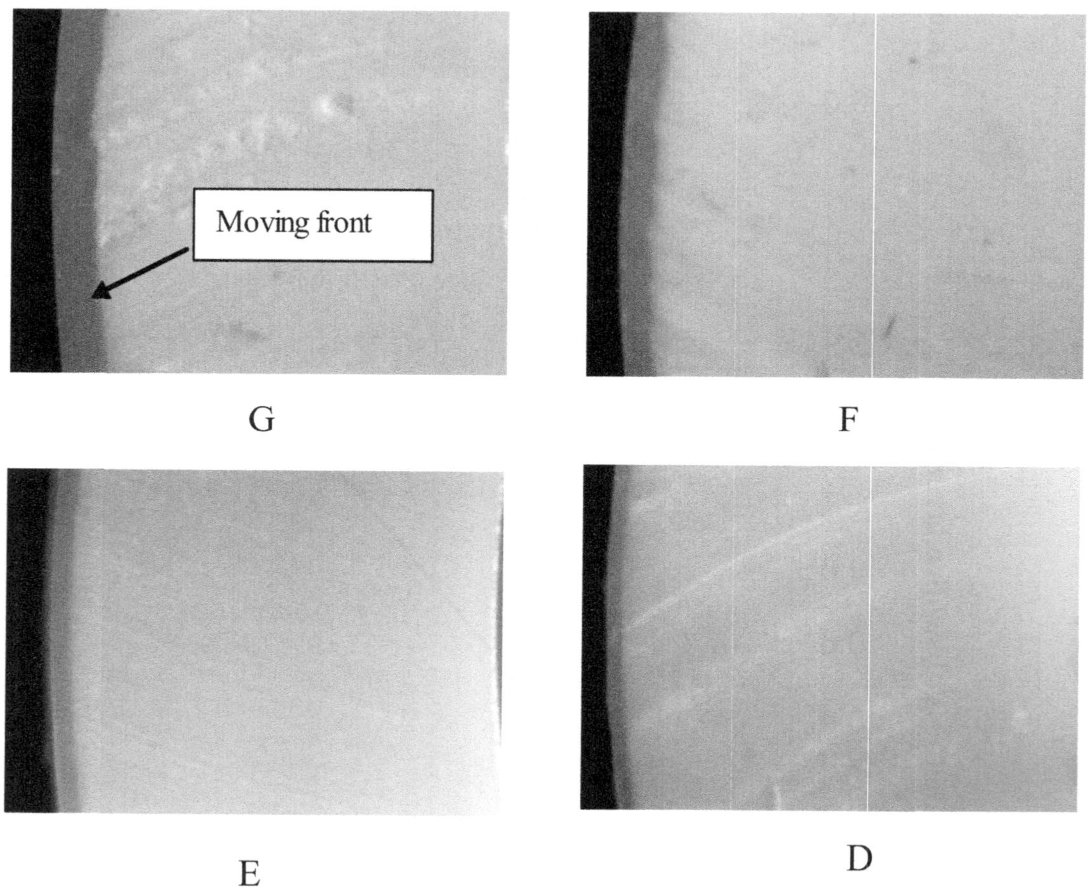

Figure 5.23 Moving front tests for PVC SP. D in mixtures of gasoline and toluene (D, E, F and G) (No moving front was observed for mixtures A, B and C) (For composition, see Table 5.3)

after one week of exposure. For similar total BTEX (v/v) in gasoline (mixtures D, E, F, G in Table 5.3), the penetration distance of moving front was strongly dependent on the fraction of toluene. Higher fractions of toluene resulted in greater penetration. Among the BTEX compounds, toluene and benzene are more aggressive than ethylbenzene and xylene to PVC pipe materials, which was confirmed by moving front tests (data not shown). The concentration of total BTEX as well as the individual concentrations of benzene and toluene in gasoline should be considered in future studies.

Comparison of Relative Susceptibility to Permeation of Pipes Obtained from Different Pipe Manufacturers

A 48-hour toluene exposure test was conducted for 58 types of PVC pipes. The pipes came from 5 different pipe manufacturers in sizes ranging from 0.5-inch to 2-inch. The swollen layer thickness ($L_{s,t}$) for each pipe sample was measured for exposure times of 6, 12, 24, 30, 36, and 48 hours. As discussed previously, $L_{s,t}$ was characterized by a square-root-of-time dependence. The regression equation, $L_{s,t} = k\ t^{1/2}$, was found to be a good fit of the data since the correlation (r^2) was greater than 0.995. The slope of the regression line, k, is defined as the growth rate of the thickness of swollen layer and was used as the index to evaluate the relative susceptibility to permeation.

The k value obtained for 58 samples ranged from 0.181 to 0.236 mm/hour$^{1/2}$. The mean and median values were 0.208 and 0.209 mm/hour$^{1/2}$, respectively, indicating a normal distribution. Although a difference in k values was detected, it was not clear whether the difference was due to measurement errors or the nature of pipe properties. To answer the question, three pipe samples (identified as Sample 8, Sample 10, and Sample 16) were selected and a 48-hour toluene exposure test was conducted in quintuplicate for each sample. The results are shown in Figure 5.24. The measurement was very reproducible and quintuplicate samples gave identical results. The differences in k values for the three samples (sample 8, sample 10 and sample 16) were statistically significant. Based on the results above, we can conclude that the moving front test is an effective tool to compare the relative susceptibility of PVC pipes to permeation.

Efforts were also made to correlate the k values with pipe dimensions (Figure 5.25). It appears that the pipe sizes used in this experiment did not significantly influence the k values.

Comparison of Relative Susceptibility to Permeation for PVC-U and PVC-M

The thickness of swollen layer ($L_{s,t}$) versus square root of time for PVC-M and PVC-U samples exposed to four solvents (chloroform, TCE, benzene and toluene) are presented in Figure 5.26. The two pipes exhibited a very similar behavior for solvent penetration. The thickness of the swollen layer was linear with the square root of time for both pipes. The penetration rates of the four solvents showed the same order in both pipes (chloroform > TCE > benzene > toluene). The k values (defined earlier) for PVC-M for each solvent were slightly lower than that for the PVC-U control specimen. The measured k for this single specimen of PVC-M (0.189 mm/hour$^{1/2}$ in toluene) was within the range of k values for the 58 specimens of PVC-U reported above, but was among the most resistant specimens tested. Future work should include more testing of PVC-M to adequately compare its resistance to permeation to that of PVC-U.

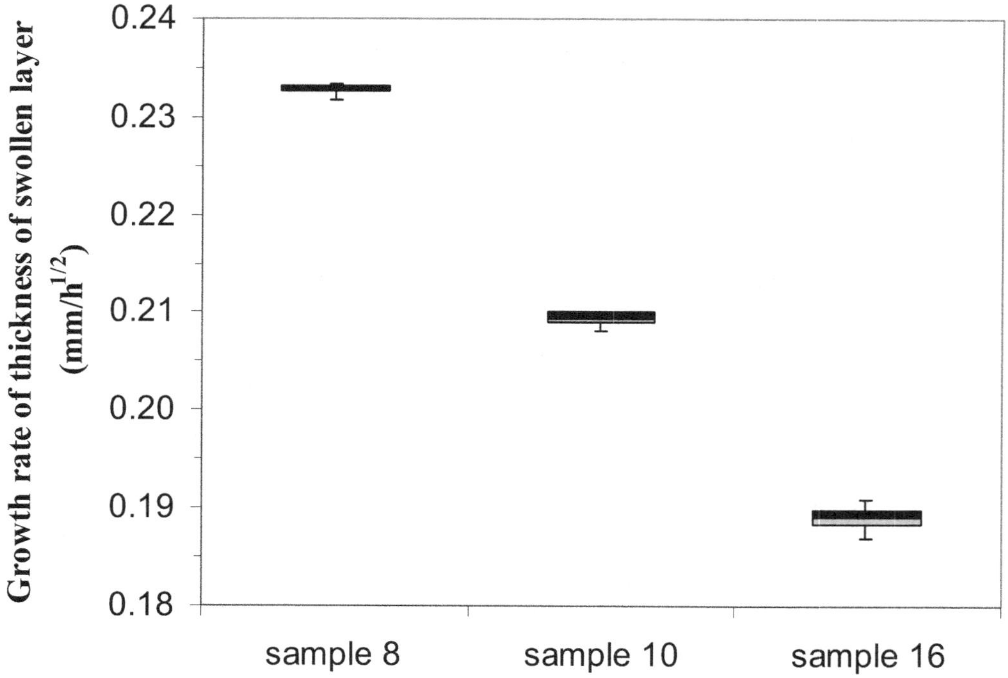

Figure 5.24 Comparison of k values obtained from exposure to toluene for Samples 8, 10 and 16

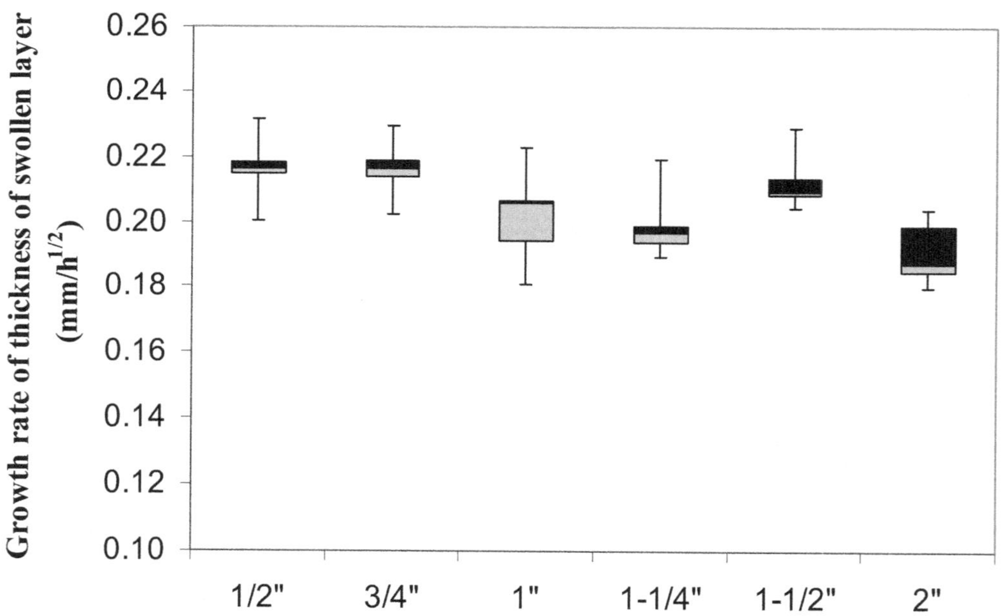

Figure 5.25 Comparison of k values obtained from exposure to toluene for different pipe sizes

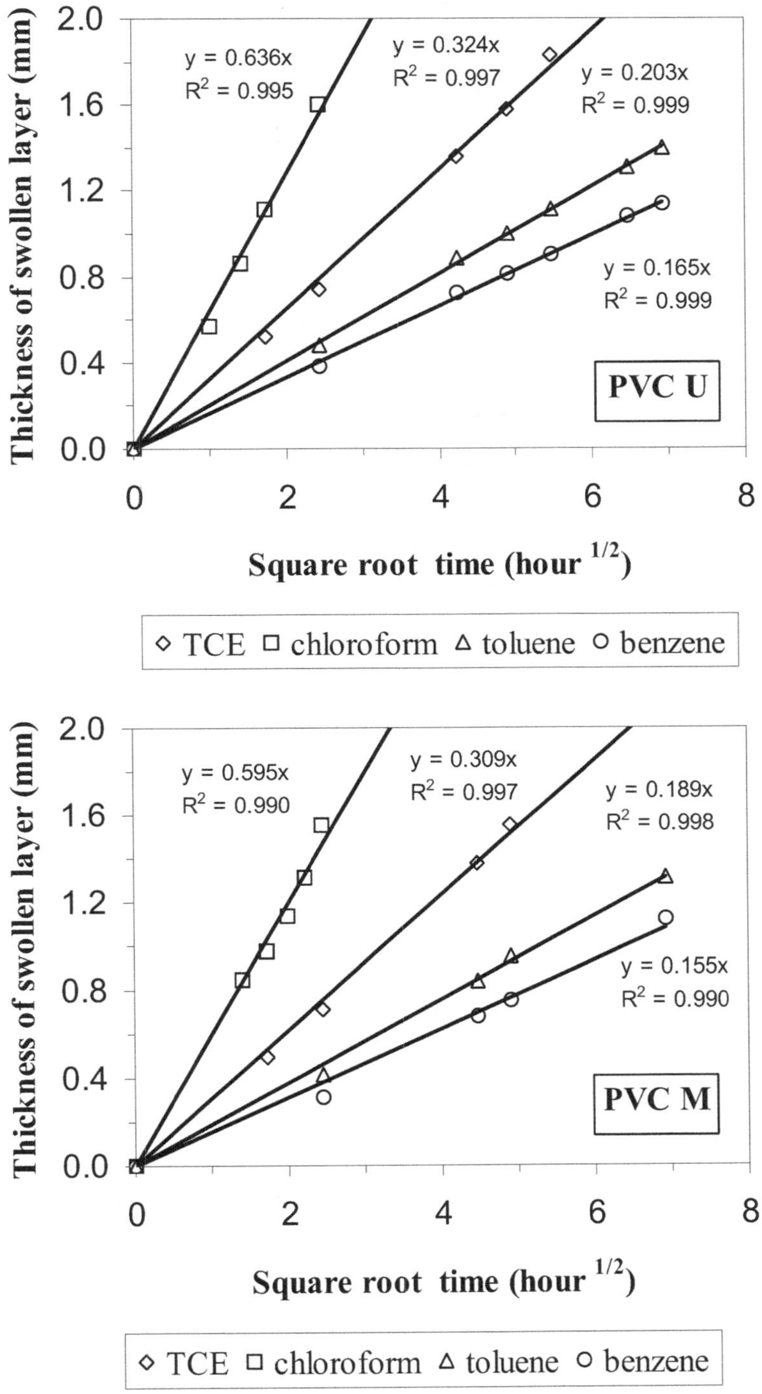

Figure 5.26 Comparison of the relative susceptibility to permeation of PVC-U and PVC-M using the moving front test

Prediction of Behavior of Organic Solvents Penetrating through PVC Pipes Based on Physiochemical Properties of the Solvent

Attempts were made to correlate the penetration rates of different organic solvents with their properties such as the octanol-water partition coefficient, the molecular diameter, and the Flory-Huggins interaction parameter. No correlations were found.

However, there is a strong direct correlation between the equilibrium weight gain and the penetration rates (Figure 5.27). Therefore, a simple sorption (weight gain) test can be used to predict whether an organic solvent is aggressive to PVC pipes or to predict the penetration rate of the solvent.

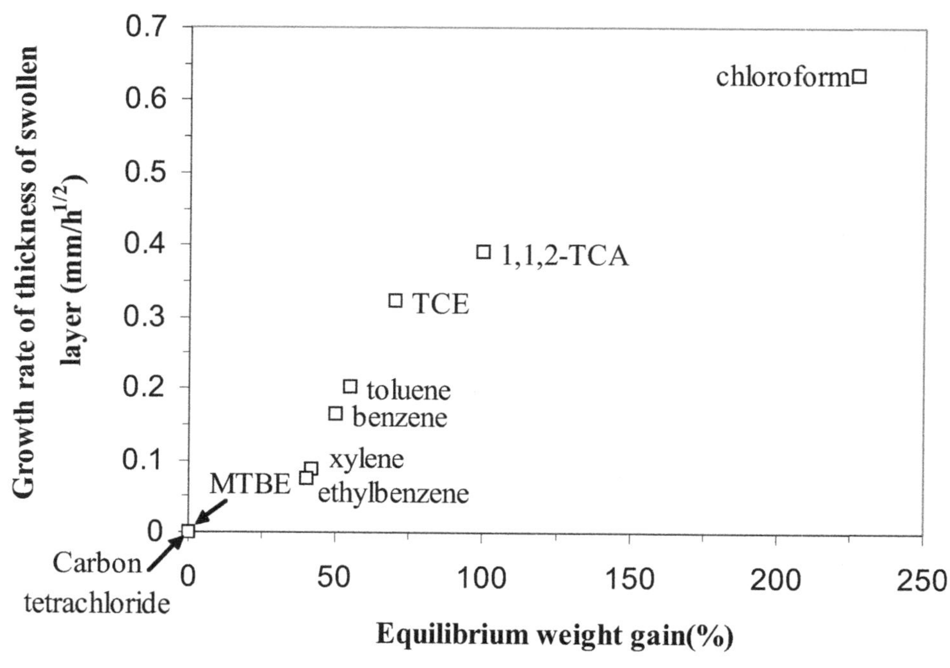

Figure 5.27 Correlation between equilibrium weight gains and growth rates of the thickness of the swollen layer

NEAR INFRARED (NIR) SPECTROSCOPY

Introduction

Near infrared (NIR) spectroscopy is a fast and non destructive technique for molecular analysis and for quantitative measurement of organic functional and molecular groups. Functional groups such as CH-, CH_2-, C=O, and O-H in organic compounds vibrate when they absorb light of a certain wavelength. By measuring the reflectance, transmittance or transflectance of light using spectrometers, the characteristic spectrum obtained can provide information on the composition and properties of the test material.

Infrared covers the electromagnetic wave frequencies from 300 gigahertz (GHz) to 400 terahertz (THz) with wavelengths from 1 mm to 750 nm (Figure 5.28). The infrared range is very broad and may be divided into three ranges: far-infrared from 300 GHz (1 mm) to 30 THz (10 μm); mid-infrared from 30 to 120 THz (10 to 2.5 μm); and near-infrared from 120 to 400 THz (2,500 to 750 nm). The near infrared region is the closest to the visible light range where the energy is higher and the wavelength is shorter than the other two ranges.

Molecular bonds of a sample can absorb or emit photons, thereby changing their energy state. When the frequency of the radiated infrared is the same as the frequency of the vibration of the molecules, a photon is absorbed and the molecule is excited, passing from its ground vibrational state to a higher state. Each molecule absorbs certain infrared light frequencies that make the dipole moment change, producing stretching vibrations that affect bond length, bending vibrations that affect bond angle, and molecular rotations (Volland 1999). Two conditions must be satisfied for the molecule to absorb infrared radiation. First, the radiation frequency must match the fundamental frequency of the molecular vibration. Second, the radiation and the molecule must interact in such a the way that the dipole of the molecule changes in the same direction as the electric field vector created as result of the radiation (Thermo Electron 2005).

When a sample is irradiated with infrared radiation, a part of this radiation is reflected (also called specular reflectance), the other portion enters the sample to a certain depth depending on the wavelength (Hruschka 1987). Once in the sample, the radiation can be transmitted through or can be reflected as diffuse reflectance (Dryden 2003). The amount of radiation reflected is usually expressed as log (1/reflectance).

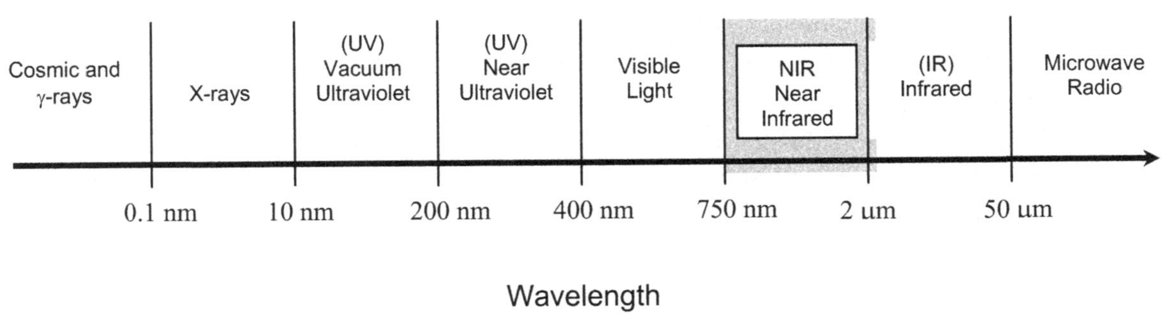

Figure 5.28 Electromagnetic spectrum

The infrared absorbance spectrum for a typical sample provides the wavelengths of infrared radiation absorbed and the percent of the incident light that passes through the molecule. This produces a curve with various peaks and valleys at different wavelengths corresponding to the absorption bands. The spectrum is composed of overtones of O-H, S-H, N-H, and C-H vibrations which are transitions of more than one energy level.

NIR light may pass through glass and optical glass fibers without modification, which allows a direct analysis of diverse materials (Choquette et al. 2002). Most samples do not need preparation and can be used repetitively as NIR radiation does not alter the structure or composition of the samples. NIR is used as a tool for quality control in polymer production and plastic recycling (Fischer and Eichhorn 1998, Rohe et al. 1998, Fischer and Pigorsch 2000, Barnes et al. 2005). It is also used to determine the concentrations of organic compounds in gasoline and other properties of fuels (Bohacs et al. 1998, Chung et al. 2001, Ozdemir 2005). Since permeability depends upon polymer chain and mobility, which amounts to vibrational bending of chemical bonds, NIR spectroscopy appeares to hold promise for assessing the susceptibility of PVC pipes to permeation. There are currently no other studies on using NIR to assess pipe permeation by organic solvents.

Chemometrics are used to analyze the spectral data. Chemometrics is the science of relating measurements made on a chemical system or process to the state of the system via application of mathematical or statistical methods. The statistical procedures in NIR chemometrics, such as principal component analysis (PCA), reduce the noise and enhance signals while retaining the relevant information by reducing the spectral data size. The principal components (PC) in PCA are the axes of the new system in which the data is projected and in which every PC represents a direction of variation. Thus, the first PC represents the largest direction of variability in the data while the second PC is orthogonal to the first one and represents the second direction with the next highest variability in the data.

The spectral data are correlated with reference data to create a regression model. Reference data such as the physicochemical properties of the material are obtained by conventional laboratory methods. NIR is laboratory dependant initially but is a fast and reliable technique once a model is created and validated.

The objective of the first study was to explore whether NIR can be used as a predictive test method by correlating the permeation susceptibility of different PVC pipes with their NIR spectra. The rate of advance of the moving front upon exposure to toluene was taken to be a measure of susceptibility of a pipe to permeation. This rate, k ($mm/h^{1/2}$), was determined using a 48-hour moving front test. The moving front studies described above have shown that the progression of the moving front (mm) was linearly proportional to the square root of time ($h^{1/2}$) for exposure to pure solvents. Differences in pipe manufacturing and addition of additives may result in spectral differences between PVC pipes which may be correlated to permeability.

The objective of the second study was to track the permeation by different organic solvents and aqueous saturated solutions of solvents. Changes in molecular bonds may be expected due to interactions of the solvents with the pipe polymer and the presence of the solute in the pipe polymer may result in spectral differences relative to nonpermeated pipe, thereby providing a way to detect the extent of permeation of the pipe.

Materials and Methods

Study 1: Susceptibility of PVC Pipes to Solvent Permeation

Pipe Selection. A group of 28 pipes was purchased from four hardware stores in central Iowa and was used for developing the calibration models. The pipes were from 5 manufacturers and consisted of 6 different diameters ranging from ½ inch to 2 ins. A second group of 30 pipes consisted of 7 different diameters (½ inch to 3 ins.) was used to validate the calibration models. Table 5.4 lists the pipes used and their sizes and manufacturers.

Laboratory Data for the Calibration. The solvent used in this study was toluene. Toluene was selected because it is a major component of gasoline and is safer to use than other gasoline compounds such as benzene. The moving front tests were conducted by immersing 1.5 cm pipe specimens in toluene and measuring the progression of the moving front over a 48-h period. Samples were removed at different times, wiped dry with a paper towel, and the progression of the moving front (in mm) was measured with an Olympus BHM reflected light microscope (Olympus America Inc., Center Valley, PA). Five optical measurements were taken equidistantly and averaged. The slope, k (mm/h $^{1/2}$), found from the plot of advance of the moving front (mm) versus the square root of time was used as a measure of the permeation susceptibility of each pipe.

NIR Scanning of Pipes. Three 5-inch long samples were cut from the purchased pipes: one sample cut from the center of the purchased pipe and the other two samples cut from the ends of the pipe. With triplicate samples of each type of pipe, there were a total of 84 samples in the first batch and 90 samples in the second. The PVC pipes (not exposed to toluene) were first cleaned with Windex to remove dirt and then dried for 24 hours. Two instruments, Foss NIRsystems 6500 (Silverspring, MD) and Labspec Pro model A108310 (Analytical Spectral Devices, Inc (ASD), Boulder, CO), were used to scan the PVC pipes.

The Foss NIRsystems 6500 measures the spectral range from 400 nm to 2,500 nm, including a portion of the visible light spectrum which is read by a silicon detector (from 400 nm to 1,100 nm). A PbS detector measures the non-visible wavelengths. A pipe sample was placed in a transport cell which moves past the light source. The instrument was programmed to scan about ¼ of the transport cell (Figure 5.29). The reflectance spectra expressed as (log (1/Reflectance)), were collected at 2 nm intervals and were analyzed statistically with The Unscrambler® v9.2 statistical package (CAMO Software, Inc., Woodbridge, NJ). The size of the transport cell limited the scanning of pipes with diameters larger than 1¼ ins. For pipes larger than 1¼ ins., pipes were cut to fit the cell before being scanned.

The Labspec Pro model A108310 (ASD) has a spectral range from 350 to 2,500 nm with a scanning time of 0.1 second, taking reflectance or transmittance data every 2 nm. The device for collecting spectra in this instrument was a fiber optic input (moglight).

Every pipe sample was scanned 6 times using the ASD instrument and the Foss instrument except for samples larger than 1¼ ins., for which the pipe samples were scanned twice by the Foss instrument only. The total number of scans for the ASD instrument and the Foss instrument were 1,044 and 828, respectively.

Chemometrics on Spectral Data. NIR data for wavelengths from 1100 nm to 2500 nm were used in the development of the calibration models. The calibration models used the raw spectral data and data that were pretreated with first and second Savisky-Golay derivatives. Savisky-Golay derivatives are mathematic algorithms for smoothing of the data by enhancing the

Table 5.4 Pipe samples information for NIR study

First Group			Second Group		
Pipe ID	Diameter (inches)	Manufacturer	Pipe ID	Diameter (inches)	Manufacturer
1	½	J-M	29	½	Genova
2	½	Suven-Line	30	½	J-M
3	½	J-M	31	½	J-M
4	½	American	32	½	N. American
5	½	Genova	33	½	Charlotte
6	¾	Genova	34	¾	J-M
7	¾	J-M	35	¾	J-M
8	¾	American	36	¾	J-M
9	¾	American	37	¾	Genova
10	¾	J-M	38	¾	J-M
11	¾	J-M	39	1	J-M
12	1	Genova	40	1	Genova
13	1	J-M	41	1	N. American
14	1	American	42	1	J-M
15	1	J-M	43	1¼	N. American
16	1	J-M	44	1¼	J-M
17	1¼	American	45	1¼	J-M
18	1¼	Genova	46	1¼	Genova
19	1¼	Suven-line	47	1¼	J-M
20	1¼	J-M	48	1½	Charlotte
21	1¼	Charlotte	49	1½	J-M
22	1½	American	50	1½	J-M
23	1½	J-M	51	1½	Genova
24	1½	J-M	52	1½	N. American
25	1½	Charlotte	53	2	Genova
26	2	J-M	54	2	N. American
27	2	J-M	55	2	J-M
28	2	American	56	2	J-M
			57	3	J-M
			58	2	J-M

relevant peaks in the spectra to make them more visible and to allow for correction of the effect of overlapping peaks (Dryden, 2003). In order to reduce the number of variables and, hence, the amount of irrelevant spectral data, the principal component analysis (PCA) method was used. The principal components represent the new axis of the system (latent vectors) which have been built following the variability direction of the data in decreasing order, from more data variability accounted in the first PC to less variability accounted in larger PCs. Using partial least squares (PLS), these new latent variables were built ensuring maximum explanation of the reference data (maximum correlation). The final equation has the form of:

$$Y = b_0 + b_1X_1 + b_2X_2 + ... + b_pX_p.$$

where b_i are the regression coefficients estimated by minimizing the residual sum of squares and X are the predictor variables.

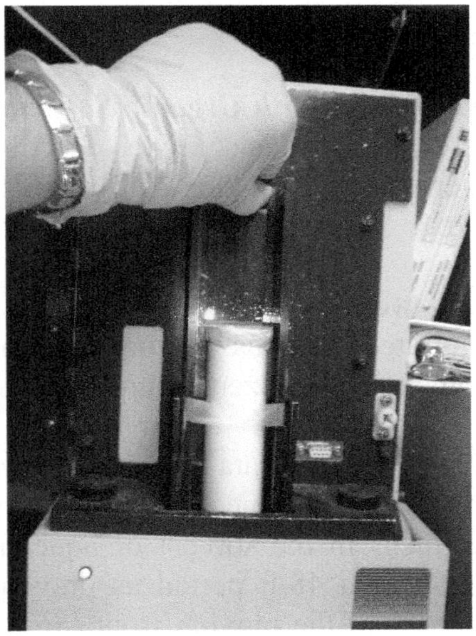

Figure 5.29 Sample in the transport cell of Foss NIRsystems 6500

Two different approaches were used to develop the calibration models. The calibration models were then validated using the data from the second batch of pipes. In the first approach, calibration models were developed using the data from the first batch of 28 pipe samples. Two models, one from each instrument, were developed and compared. For easy identification, models developed using this data usage approach were given the designation "M1."

The second approach used to develop PLS independent models was by using ¾ of the whole data set for calibration and then using the remaining ¼ of the data set to validate the PLS models developed. These models were designed as "M2."

Finally, locally weighted regression (LWR) models were used since LWR has been shown to be a useful technique for developing calibration models for heterogeneous or cluster-based data (Centner et al. 2000). The models were built with selected samples of the local principal component regression (PCR). PCR follows the same procedure as PLS but the selection of latent variables is only according to data variability and does not account for the dependent variable–latent variables correlation (Reich 2005). LWR models were developed for ¾ of the data set using Foss6500 instrument while the remaining ¼ was used for validation. The resultant models were designated as "M3."

The calibration models were described by the coefficient of determination (R^2). The accuracy of the validation was reported using the standard error of prediction (SEP) and the ratio of the standard error of prediction to the reference data deviation. This ratio is the relative predictive determinant (RPD) and indicates the predictive accuracy of the model. Acceptable RPD values for validation are above 3 while excellent accuracy values for validation are above 8 (Williams and Norris 2001).

Study 2: Permeation Tracking with NIR

Selection and Treatment of Pipes with Organic Solvents. Three 1-inch diameter PVC pipes (PVC SP. C, D, and I) were used for this study (as described above). The manufacturers for PVC SP. C, D and I were Charlotte Pipe and Foundry Co., Livingston, NJ, J-M Manufacturing Company, Monroe, NC and Cresline Plastic Pipe Company, Evansville, IN, respectively.

The exposure media used were pure solvent (premium gasoline, toluene or benzene) 100%, 80%, 40% and 20% saturated aqueous solutions of toluene or benzene prepared as described in Chapter 3. Pure toluene and benzene (purities > 99%) were obtained from Fisher Scientific, Chicago, IL.

Laboratory data for the calibration. Data from the moving front test and the equilibrium sorption test were used as the laboratory reference data. The moving front tests were conducted by immersing 1.5 cm pipe specimens in the solvent or aqueous solutions and measuring the progression of the moving front over a 48-h period as previously described. The slopes, k ($mm/h^{1/2}$), of the plot of the thickness of the moving front (mm) versus the square root of time were used in the calibration.

For the sorption tests, 1.5 cm PVC samples were immersed in solvents or aqueous saturated solutions in 120 mL I-Chem jars with Teflon®-lined caps (Chase Scientific Glass Inc., Rockwood, TN). After wiping with a paper towel and air drying for 30 seconds, the samples were weighed. The data collected were percent weight gain or weight gained per unit pipe length (g/cm). Table 5.5 summarizes the experimental matrix and the length of time for each experiment. For 40% and 20% saturated aqueous solutions there was no measurable moving front or weight gain, and the number of days of immersion was used for the calibration models.

NIR scanning of pipes. The pipe samples (about 13 cm (5 ins.) long) were washed with water and Windex and dried for 24 hours at room conditions. The ends of the PVC samples were sealed with a glass slide on both sides with Loctite® epoxy putty (Henkel technologies, Rocky Hill, CT). Before the pipe was completely sealed, the pipe was filled with water to about ½ inch from the top (Figure 5.30). Duplicate samples were set up for each solvent at each concentration. PVC samples were vertically immersed in individual 1-L I-Chem Teflon®-capped jars filled with solvent or aqueous solutions to minimize headspace (Figure 5.31). Aqueous solutions were transferred to each jar by using a Masteflex pump and Teflon tubes, and were replaced every week to ensure constant concentration during the study.

The samples were taken out of the jars, wiped dry and scanned with the Foss NIRsystems 6500. Samples in gasoline, benzene and toluene were scanned every day (6 scans per day) for 9 days until close to solvent breakthrough into the pipe-water. Samples in aqueous solutions were scanned once every two weeks with 4 scans for every sample over a 6-month period.

Chemometrics on spectral data. Spectral data were taken for wavelengths from 1,100 nm to 2,500 nm. This range included the first and second CH overtone regions (around 1,650 – 1,790 nm and 1,150 – 1,250 nm, respectively) and CH combination regions (around 2,200 – 2,450 nm). The third overtone region (around 850 – 950 nm) was excluded because it did not add useful information. CH absorption regions were expected to provide most of the information.

Model calibrations were developed with The Unscrambler® v9.5 statistical package (Camotechnologies, Trondheim, Norway). Raw spectral data were analyzed first, and then pre-treated with first and second Savisky-Golay derivatives using 5 points of smoothing.

Table 5.5
Test matrix for permeation tracking study

Pipes	Solvent	Dilution	Sample*	Total data points	Scans/day (reps)	Total immersion time
PVC SP. D, I, C	Gasoline	Pure	1	19	6	9 months
			2	19	6	9 months
	Toluene	Pure	3	9	6	9 days
			4	9	6	9 days
		100% saturated solution (ss)	5	14	4	7 months
			6	14	4	7 months
		80% ss	7	14	4	7 months
			8	14	4	7 months
		40% ss	9	14	4	7 months
			10	14	4	7 months
		20% ss	11	14	4	7 months
			12	14	4	7 months
	Benzene	Pure	13	9	6	9 days
			14	9	6	9 days
		100% saturated solution (ss)	15	14	4	7 months
			16	14	4	7 months
		80% ss	17	14	4	7 months
			18	14	4	7 months
		40% ss	19	14	4	7 months
			20	14	4	7 months
		20% ss	21	14	4	14 days
			22	14	4	14 days

* Triplicate samples (total = 22 x 3 = 66)

Nine PLS models were developed for pipes immersed in premium gasoline, benzene or toluene by using: (i) weight gain (sorption) in grams per cm length; (ii) percent weight gain (sorption); and (iii) the rate of progression of the moving front (mm/hour$^{1/2}$).

For aqueous solutions of toluene and benzene, PLS models were developed for PVC D only by using sorption test data (weight gain per length of pipe (g/cm)). Since the weight gain or the permeation for the 40% and 20% saturated solutions was negligible or not detectable, the time of exposure (days) was used to correlate to the 6 months of spectral data for these.

Cross validation of the models was conducted by selecting 5 to 10 samples randomly from the whole sample set. The number of cross validation samples depended on the size of the data set.

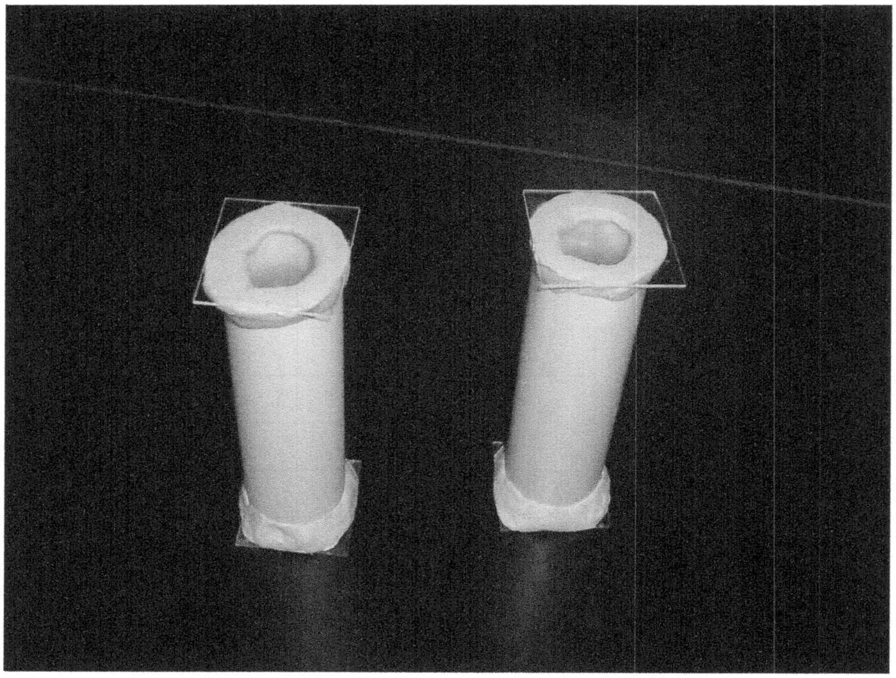

Figure 5.30 Pipe samples sealed with glass slides and epoxy

Figure 5.31 Samples immersed in jars with Teflon®-lined caps

Model best fit was reported using the following statistical parameters: coefficient of determination (R^2), standard error of the cross validation (SECV), and relative predictive determinant (RPD).

Results and Discussion

Typical NIR spectra (reported as log (1/Reflectance)) versus the NIR wavelengths for unexposed pipes and pipes exposed to toluene are shown in Figure 5.32. Of importance to Study 1 in Figure 5.32 are the NIR spectra without exposure to toluene. Subsequent exposure to toluene resulted in a change in the NIR spectra at specific wavelengths which were used in Study 2. Figures 5.33 and 5.34 show the regression coefficients for the PLS models using the PCA approach for PVC pipes exposed to toluene and benzene, respectively. The peaks and valleys for the different wavelengths in these figures correspond to various molecular groups. The higher absolute values of the Y-axis meant that more weights were placed on these wavelengths in the calibration models. Based on the model building approach discussed above, various calibration models were developed.

Study 1: Susceptibility of PVC pipes to Solvent Permeation

Calibration models (M1) using the first group of 28 pipe samples showed strong correlation of the spectra to the slopes, k (mm/h$^{1/2}$), for both raw and pre-treated spectral data using the Foss 6500 instrument. This is indicated by the R^2 as presented in Table 5.6. However, the R^2 for calibration models using the ASD instrument were lower than those using the Foss 6500 instrument. Spectral data from the ASD instrument showed a large presence of noise and irregularities and may be one of the reasons why the models for the ASD instrument were not as good as those for the Foss 6500 instrument. Treatment of the data with derivatives provided slightly poorer calibration models. Unfortunately, the M1 calibration models were unable to satisfactorily predict the data from the second group of pipe samples. The RPDs were approximately 1 while a RPD of 3 is required for an acceptable calibration model.

The second calibration model (M2) using ¾ of the whole data set and the other ¼ of the data set for validation is presented in Table 5.7. This is a common method for validation of NIR calibrations. Table 5.7 showed that the R^2 for the Foss 6500 and ASD instruments were slightly worse than that of M1 calibration models but the RPDs for the Foss 6500 instrument were slightly better than that of the M1 calibration models. Treatment of the Foss 6500 data with the first and second derivatives gave slightly better R^2 values. For the ASD instrument only, the second derivative gave better R^2 values than the raw data model. A possible reason is the clustering effects detected in the principal component analysis (see Figure 5.35) which did not allow for more accurate models.

As a check on the laboratory data, the standard error of the laboratory experiments (SEL) was estimated for 3 pipe samples with 5 repetitions each. This SEL gave a value of 0.0018 mm/h$^{1/2}$ indicating that laboratory reproducibility was not limiting the accuracy of the models.

Using LWR approach, the modeling results (M3) improved with an R^2 of 99.5% for the raw data from Foss 6500 instrument. The RPDs for the LWR models were approximately 7.3 for the raw data (see Table 5.8) while the second derivative treatment of data gave RPDs between 4.3 and 5.3.

NIR can be used to assess the susceptibility of PVC pipes to permeation by using the slope, k (mm/$h^{1/2}$), from the moving front test as a variable for calibration development and by using the appropriate statistical model. The LWR model seemed to perform the best since the approach helped to account for data heterogeneity. Since the models developed were very specific to the NIR instrument and on the data used, prediction of permeation susceptibility for pipes that do not group into the clusters identified in this calibration database (Figure 5.35) can not be assured. Adding more pipes to the database may provide increased robustness to the models for identifying additional clusters and is required if a general model is to be arrived at. However, the LWR technique used in this study can provide an indication of how well an individual pipe fits within the known PCA clusters, thereby, providing a degree of usefulness for the calibration model developed here.

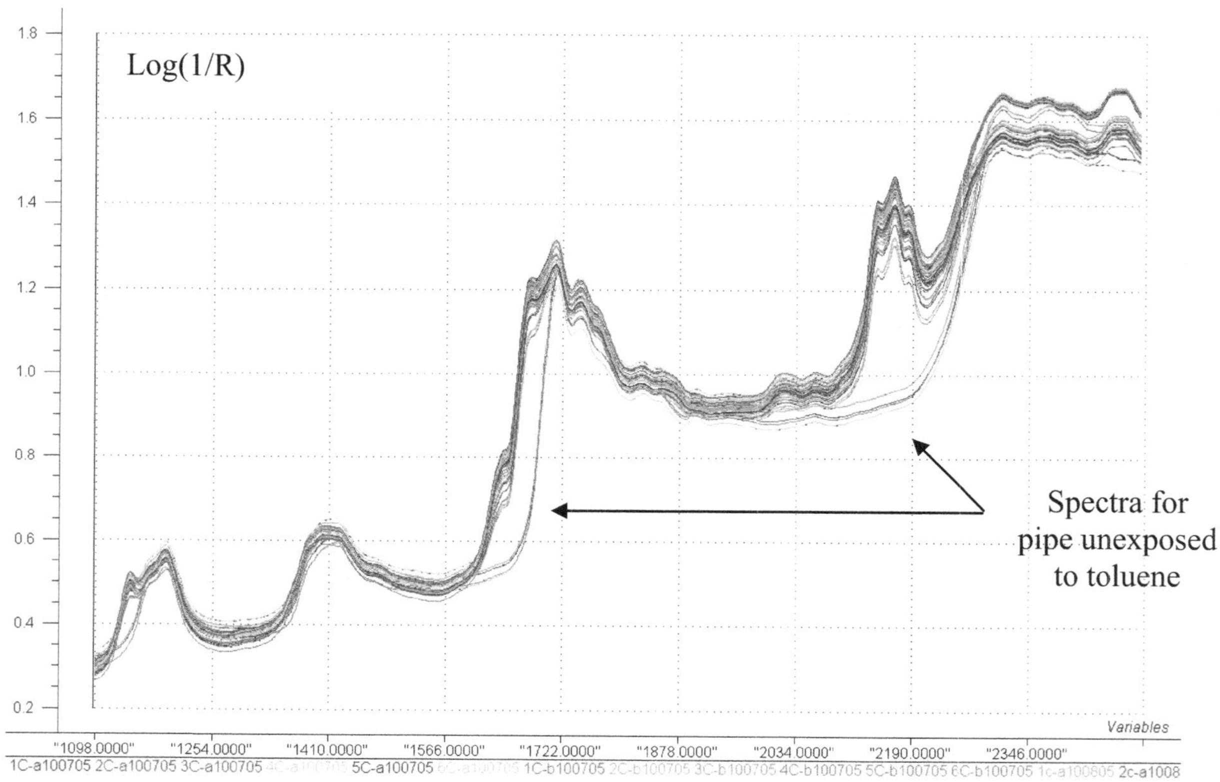

Figure 5.32 Spectral changes in PVC pipe immersed in toluene for 9 days

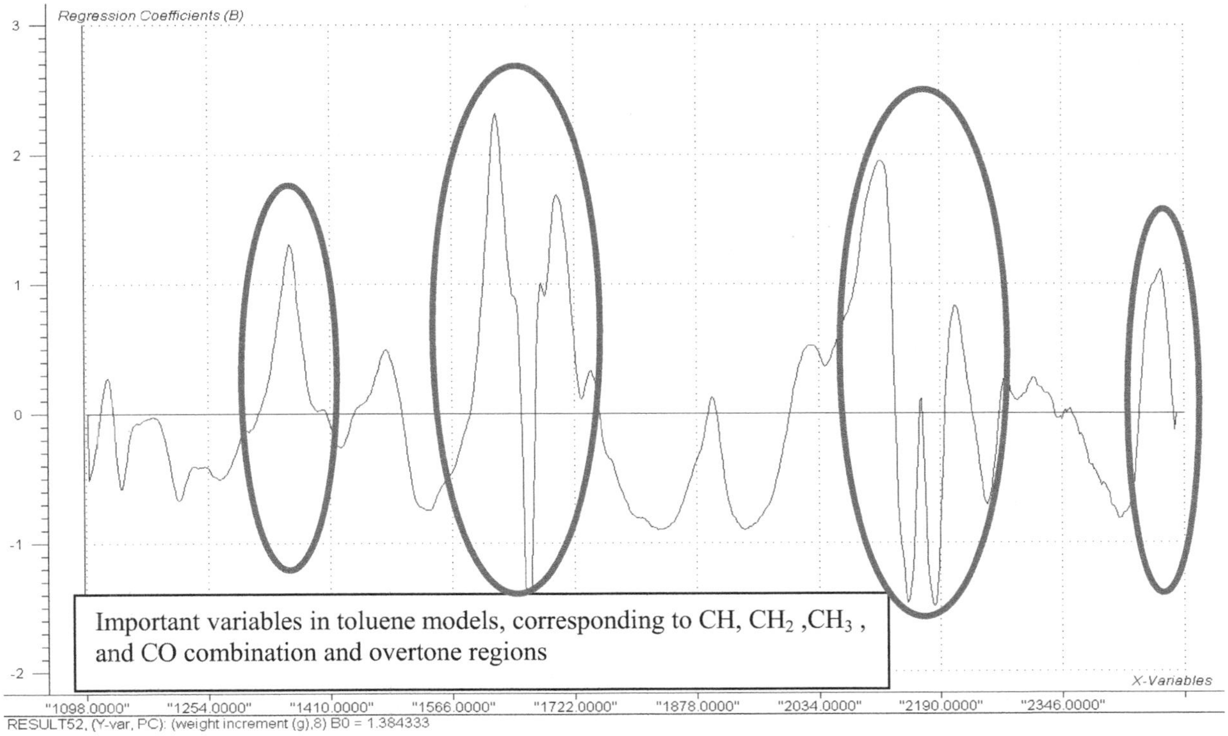

Figure 5.33 Relevant regression coefficients in model development for PVC in toluene

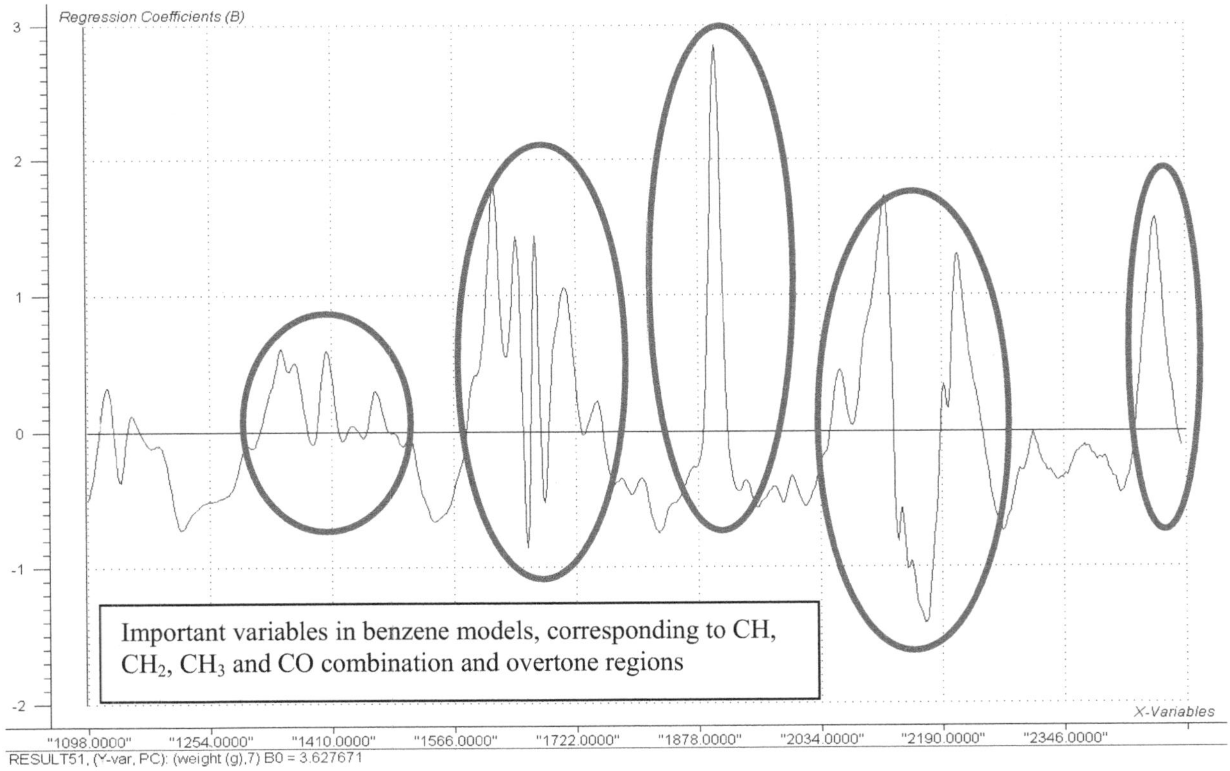

Figure 5.34 Relevant regression coefficients in model development for PVC in benzene

Table 5.6
Calibration models (M1) using two different groups of PVC pipes

Instrument	Data Treatment	M1 Calibration models		Validation with Second Batch	
		PCs*	R^2 (%)	RPD	SEP
Foss 6500	Raw data	19	95.70	1.02	0.016
	1st derivative	16	93.06	1.04	0.014
	2nd derivative	12	89.06	1.15	0.012
ASD	Raw data	15	81.10	1.15	0.015
	1st derivative	5	73.98	1.01	0.014
	2nd derivative	6	76.36	1.03	0.015

* PC – principal components

Table 5.7
Calibration models (M2) by combing the data set and using ¾ of the data for calibration

Instrument	Data Treatment	Calibration models M2 (¾ data)		Validation (¼ data)	
		PCs	R^2 (%)	RPD	SEP
Foss 6500	Raw data	18	82.19	2.33	0.0059
	1st derivative	16	87.49	2.21	0.0062
	2nd derivative	13	85.61	1.87	0.0074
ASD	Raw data	12	39.06	1.25	0.0108
	1st derivative	3	35.08	1.18	0.0114
	2nd derivative	6	51.55	1.28	0.0106

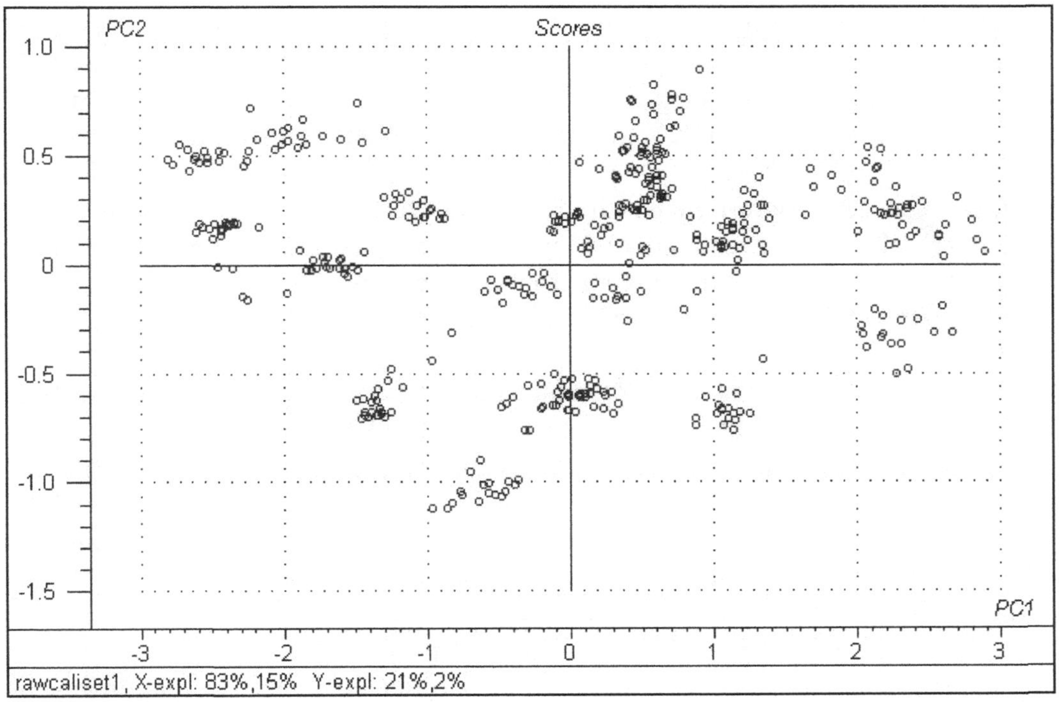

Figure 5.35 Clustering in the principal component analysis due to pipe types (PC1 and PC2 - principal component 1 and principal component 2, respectively)

Table 5.8
Calibration models (M3) using the LWR approach for Foss 6500 Instrument data (¾ of the data used for calibration, the rest for validation)

Data Treatment	Spectral Neighbors	Calibration models M3 (¾ data)		Validation (¼ data)	
		PCs	R^2 (%)	RPD	SEP
Raw data	25	7	99.5	7.30	0.0019
	50	10	99.5	7.29	0.0019
	75	12	99.6	7.35	0.0018
1st Derivative	25	7	99.5	6.25	0.0022
	50	11	99.5	5.84	0.0023
	75	12	99.4	6.09	0.0022
2nd Derivative	25	5	99.1	5.33	0.0026
	50	8	98.8	4.70	0.0029
	75	8	98.3	4.30	0.0032

- Spectral Neighbors – defined as the number of samples chosen as similar to the unknown sample for use in developing the model for the unknown.

Study 2: Permeation Tracking with NIR

Models for pure toluene and benzene studies. The NIR spectral data for pipes exposed to toluene and benzene for different periods of time were correlated with three laboratory reference data sets: (i) weight gain in grams per cm length; (ii) percent weight gain; and (iii) the rate of progression of the moving front (mm/hour$^{1/2}$). The PLS models statistics are summarized in Table 5.9.

For PVC pipes exposed to pure toluene, the R^2 values were equal to 98% and the RPD values in the range of 6 – 10 for all treatments of the data (raw, first and second derivatives), indicating good to excellent accuracy. The moving front test appeared to be the best laboratory reference data for all treatments of the spectral data with the first derivative being the best data treatment (RPD = 9.76).

For PVC pipes exposed to benzene, the R^2 values ranged from 97% to 98% and the RPD values were 5 to 6, indicating acceptable predictive accuracy. The best predictive accuracy was achieved with the moving front test using raw data treatment (RPD = 9.69), although this experiment was based on only 7 data points.

Models for gasoline. The NIR spectra over 9 months of immersion in free product gasoline are shown in Figure 5.36. In addition, the moving front progression and weight gained by sorption were negligible. In this case, the spectral data was correlated with the number of days the pipes were immersed in gasoline. The PLS calibration model using raw data showed a very strong linear relationship with days of exposure, also reflected in the validation results (Figure 5.37), which indicates that changes in the pipes were linearly related with time for all pipe types in the model. Although neither moving fronts nor significant % weight gains (sorption) were observed, changes in the outer surface of the pipe due to exposure to gasoline compounds apparently occurred as a function of time (R^2= 99%, SECV = 9.27 days, and RPD = 8.79).

Models for aqueous solutions of benzene and toluene. Both 40% and 20% aqueous solutions of toluene and benzene gave very low or no permeation in the PVC pipes and no moving fronts were detected. Calibration models were developed for the 80% and 100% aqueous saturated solutions using the sorption data only. The moving front test could not be used as reference data for these two concentrations because the moving fronts took several weeks to develop and the data would not have correlated well with the NIR data, which were influenced by permeation much sooner. For aqueous solutions of toluene (100% and 80%), both raw and first derivative pretreatment of the data gave the best models with R^2= 97% and RPD above 6 for sorption data (% weight gain) (Table 5.9). For aqueous solutions of benzene (100% and 80%), the second derivative data gave the best model with R^2=96% and RPD above 6 for sorption data (% weight gain).

Since the weight gain or the moving front were negligible or low for the 40% and 20% saturated aqueous solutions, the time of exposure was used as the laboratory reference data. Using time as a reference data allowed tracking of the permeation process from preliminary stages since NIR was able to detect changes in pipes during the first two weeks of permeation. The calibration models are presented in Table 5.10. For all models, the R^2 were greater than 94% while the RPD ranged from 3.7 to 7.7. However, the RPDs appeared to be lower for less saturated solutions of toluene and benzene.

Table 5.9
Modeling of permeation of PVC pipes with various solvents and aqueous saturated solutions

Solvent/ Solution	Reference Data	Pretreatment	R^2 (%)	RPD	SECV*	PCs	SEL
Pure toluene	Moving front (mm/hour$^{1/2}$)	Raw data	98.4	7.98	0.14	8	0.13
		1st derivative	99.0	10.03	0.11	6	
		2nd derivative	98.7	9.56	0.12	6	
	Sorption test (g/cm pipe)	Raw data	98.1	7.21	0.17	8	0.05
		1st derivative	98.6	8.58	0.14	6	
		2nd derivative	98.4	8.48	0.15	6	
	Sorption test (% weight gain)	Raw data	98.1	7.36	2.27	8	0.26
		1st derivative	98.6	8.72	1.92	6	
		2nd derivative	98.6	9.52	2.05	6	
Pure benzene	Moving front** (mm/hour$^{1/2}$)	Raw data	99.1	9.69	0.08	9	0.015
		1st derivative	98.9	10.31	0.08	7	
		2nd derivative	98.8	10.32	0.09	7	
	Sorption test (g/cm pipe)	Raw data	98.4	7.83	0.14	9	0.31
		1st derivative	97.9	6.94	0.16	6	
		2nd derivative	97.6	6.71	0.17	5	
	Sorption test (% weight gain)	Raw data	98.4	8.06	1.82	9	0.22
		1st derivative	97.9	7.14	2.03	6	
		2nd derivative	97.6	6.67	2.19	5	
Toluene aqueous solution (100% and 80%)	Sorption test (g/cm pipe)	Raw data	95.2	4.78	0.04	10	n/a
		1st derivative	95.0	4.84	0.04	8	
		2nd derivative	95.2	4.52	0.04	7	
	Sorption test (% weight gain)	Raw data	97.3	6.04	0.62	11	
		1st derivative	97.2	6.48	0.63	7	
		2nd derivative	96.3	8.27	0.73	7	

(continued)

Table 5.9 (Cont'd.)
Modeling of permeation of PVC pipes with various solvents and aqueous saturated solutions

Benzene aqueous solution (100% and 80%)	Sorption test (g/cm pipe)	Raw data	95.1	4.52	0.11	10	n/a
		1st derivative	96.2	5.12	0.09	9	
		2nd derivative	95.9	4.96	0.09	8	
	Sorption test (% weight gain)	Raw data	94.9	4.53	1.51	10	
		1st derivative	95.9	5.21	1.37	9	
		2nd derivative	96.0	6.25	1.37	9	

* Units for SECV are the same as for the reference data.
** Models using 7 data points

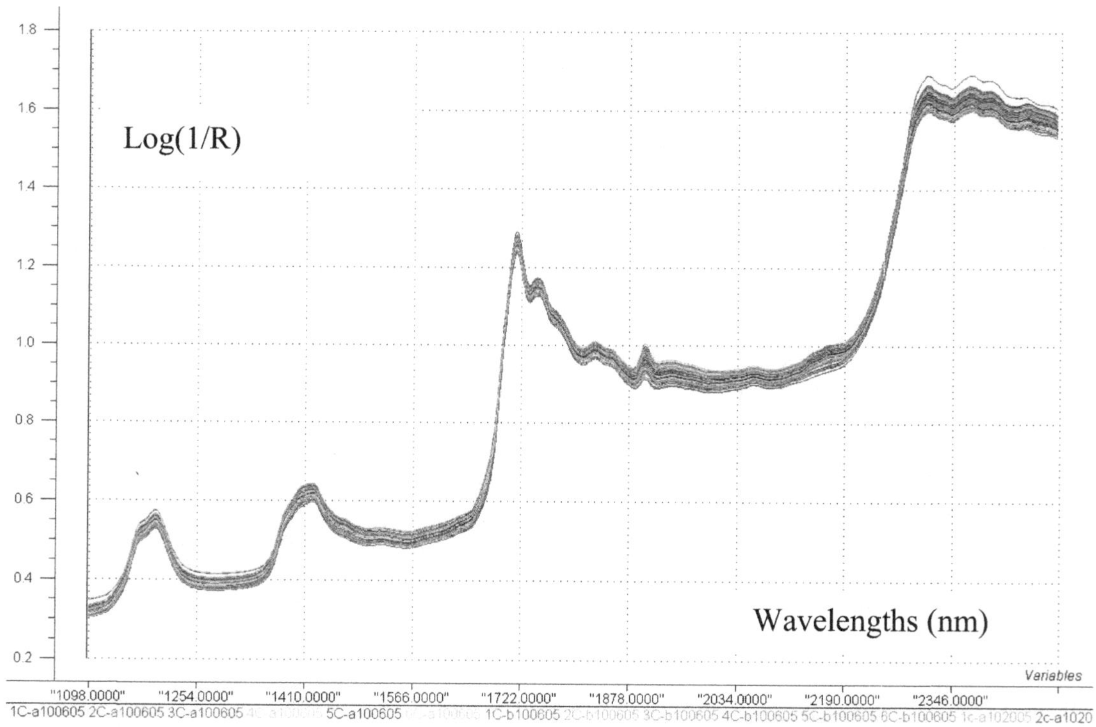

Figure 5.36 Spectra for PVC C pipe in gasoline over 9 months

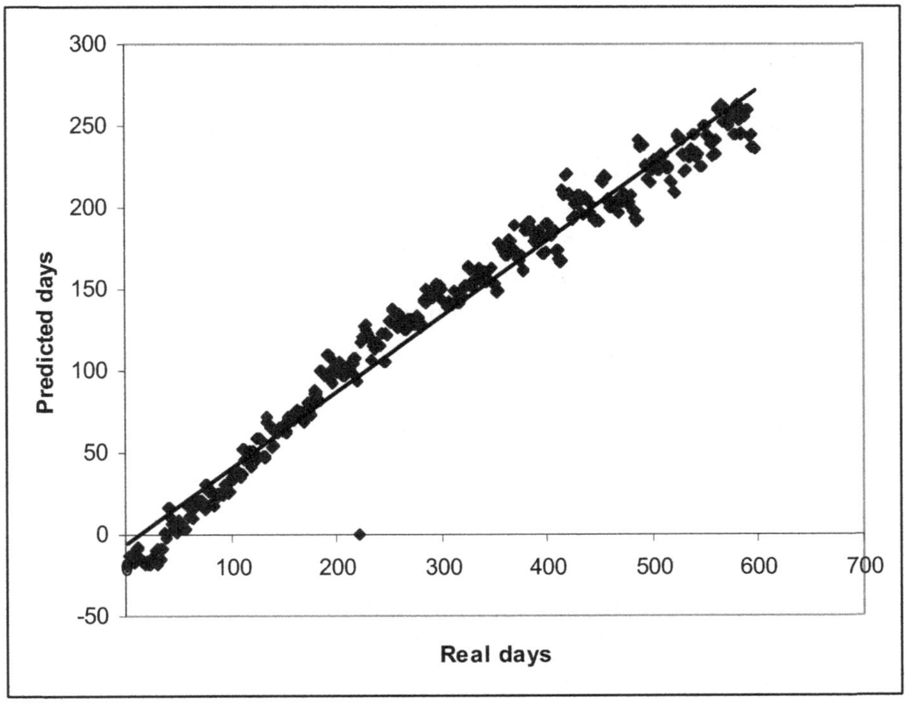

Figure 5.37 Prediction of model for PVC pipes soaked in gasoline by using days of exposure as the laboratory reference data

Table 5.10
Summary of statistics from individual aqueous saturated solutions (NIR raw spectral data calibrated with time of exposure)
Exposure time was 7 months except as noted.

Solvent	Solution	R^2 (%)	RPD	SECV (days)	PCs
Toluene	100% saturated solution	97	5.97	10.79	6
	80% saturated solution	98	7.73	8.74	8
	40% saturated solution	94	4.19	16.38	9
	20% saturated solution*	94	3.77	17.76	11
Benzene	100% saturated solution	98	7.07	8.85	8
	80% saturated solution	96	4.77	13.44	8
	40% saturated solution	96	4.85	13.65	15
	20% saturated solution*	96	4.48	15.15	11

* 14 day exposure time

Conclusions and Implications for Future Use

Study 1: Susceptibility of PVC pipes to solvent permeation

The susceptibility of PVC pipes to permeation could not be modeled with success with the global partial least squares (PLS) approach due to data heterogeneity and clustering. However, the locally weighted regression (LWR) model was able to deal with the heterogeneity of the data with an RPD of approximately 7.2 instead of 2 for PLS models. This indicates that NIR spectroscopy is capable of predicting the permeability of PVC pipes using models such as locally weighted regression (LWR), but only when the samples being tested fall within one of the clusters in the data set used to develop the calibration model. Testing more known pipe specimens will enlarge the data set and increase the probability that an unknown specimen will fall within one of the data clusters in the LWR calibration model, thereby making the calibration model more robust.

Study 2: Permeation Tracking with NIR

NIR spectroscopy is also a suitable tool for tracking the permeation process in PVC pipes. Models developed for benzene and toluene had RPDs between 5 and 9 making them suitable for screening the extent of permeation. The three laboratory reference data sets, percent weight gain, weight gain per unit length and progression of the moving front were found to be suitable for the development of calibration models. First Savinsky-Golay derivative appeared to improve the results in most individual models.

Although permeation of gasoline into PVC pipes was negligible, the NIR spectra were found to correlate with the days of exposure to gasoline. Similarly, the number of days of exposure to aqueous saturated solutions of toluene and benzene were found to correlate with the NIR spectra. However, the lower the aqueous saturation concentration, the higher was the prediction error.

The above results imply that NIR reflectance spectroscopy can be used as a predictive test tool provided a database of reference laboratory data is established for a given NIR spectrometer with updating of the calibration set when new types of pipes are used or introduced. Predictions using the model, however, are limited to the types of PVC pipes within the database and may not be able to predict with certainty for the types of PVC pipes not in the data base.

The results also indicate that NIR reflectance spectroscopy may be used as a test method for developing information on the exposure history of PVC pipe, such as discerning longer from shorter periods of exposure to gasoline.

SUMMARY

This chapter describes three predictive test methods to predict the susceptibility of PVC pipe to permeation. The accelerated techniques were: (i) gravimetric sorption; (ii) moving front test (visualization); and (iii) NIR reflectance spectroscopy.

Weight gain of PVC pipe specimens in the gravimetric sorption test can provide information on the susceptibility of PVC pipes to various solvents and aqueous solutions. The sorption test is a fast and easy method but provides only indirect information on the extent of solvent permeation and permeation rates, and may not be suitable or accurate for small mass changes. The moving front test is a fast and accurate method for visualizing the extent of solvent permeation into the pipe using a light microscope. Direct measurement of the advance of the moving front enables prediction of solvent breakthrough time as well as prediction of the level of solvent activity required to initiate a moving front.

Equilibrium sorption and moving front tests showed that PVC pipe is not permeated by exposure to water that is ≤ 60% saturated with benzene, toluene, or TCE (single contaminant experiments). Those tests also showed that PVC pipe is not permeated by exposure to ≤ 40% (v/v) benzene or toluene in NIST reference fuel (single contaminant experiments) or by exposure to ≤ 40% (v/v) of an equal-volume mixture of benzene, toluene, ethylbenzene and xylene in NIST reference fuel, whereas the equivalent experiments at ≥ 60% (v/v) do cause permeation, as evidenced by sorption and initiation of moving fronts. Results from moving front tests with gasoline fortified with toluene were consistent with the observations in the reference fuel experiments that a combination of BTEX compounds exceeding 40% (v/v) in gasoline is required to initiate swelling and to cause permeation of PVC pipe. All of this evidence suggests that 1) the impact on PVC pipe of the activities of the individual BTEX compounds in an exposure medium is additive and 2) the performance of PVC pipe will not be impaired upon exposure to groundwater in which the sum of the individual activities of BTEX compounds does not exceed 0.40. Contamination of groundwater by gasolines commonly available for motor fuel will not cause the sum of the activities of the BTEX compounds to exceed 0.40 and will not cause permeation of PVC pipe.

NIR reflectance spectroscopy can be used as a predictive test tool but requires laboratory reference data such as the moving front test or sorption test to build a database. Time of exposure to contaminants may also be used as reference data. Calibration models developed are specific to a given NIR spectrometer model and are only suitable for predicting susceptibility for types of PVC pipe within the database. While NIR reflectance spectroscopy is a fast and reliable tool, development of the database requires time and effort and the calibration models are not transferable to other models of NIR instruments. NIR reflectance spectroscopy may also be useful for developing information about the length of time PVE pipe has been exposed to gasoline and other hydrocarbons.

All three predictive tests can be used to rapidly test the suitability of PVC pipes for various contamination conditions. Each test has their advantages and disadvantages. If a microscope is easily available, the moving front test would be preferable since it involves a direct measurement of permeation.

CHAPTER 6
GASKETS – LABORATORY STUDIES UNDER SIMULATED ENVIRONMENTAL CONDITIONS

OVERVIEW

According to a national survey by Thompson and Jenkins (1987), a very large majority of the drinking water permeation incidents in U.S. were related to petroleum products in contact with plastic pipes such as polyethylene (PE), polybutylene (PB), and polyvinyl chloride (PVC). As a result, many studies have been conducted to understand the permeation of organic compounds in contaminated soils through plastic pipes (Berens 1985; Vonk 1986; Park et al. 1991; Holsen et al. 1991a, 1991b). Results showed that PE and PB pipes are susceptible to permeation. Therefore, copper or ductile iron pipes have been recommended and installed in gasoline contaminated areas, close to gas stations and industrial areas.

One percent of the reported permeation incidents were due to permeation through gasket materials as reported by Holsen et al. (1991a). Styrene-butadiene rubber (SBR) is the most common synthetic rubber used for pipe gaskets (more than 90%) due to its low cost. Nitrile (NBR), ethylene propylene diene monomer (EPDM), neoprene (CR), and fluoroelastomer rubber (FKM) are used for specific environmental applications. However, NBR and FKM are recommended by the industry for use in contaminated soils since the materials have good resistance to gasoline. Studies conducted by Park et al. (1991) suggested that the SBR gaskets typically used in water main distribution system, were susceptible to permeation by organic contaminants such as gasoline and toluene. However, the low number of incidents reported for gaskets may be due to the high water flow rates and lack of water stagnation, resulting in dilution to concentrations not exceeding the taste or odor limits (Holsen et al. 1991a). The common flow velocities in water distribution systems range from 2 to 10 feet per second (ft/s), which give flow rates of approximately 80 to 400 gallon per minute (gpm) in 4-inch DI pipes. Nevertheless, such contamination, although not exceeding maximum contaminant levels (MCL) set by the federal government for drinking water, may have a harmful effects on human health that have not yet been identified by epidemiological studies.

Previous studies have discussed the risk of hydrocarbon permeation through gaskets when exposed to contaminated soils and groundwater. However, there are several significant issues that need to be answered for the water industry, especially the water distribution sector. These issues include the impact of environmental conditions, typical environmental contamination levels, the characteristics of actual gasket materials, and the impact of hydrostatic pressure on permeation of hydrocarbons through the gaskets.

This study strives to understand permeation of gasoline through gasket materials and the thresholds of environmental contamination that will result in exceeding contaminant MCLs.

MATERIALS AND METHODS

Materials

Five different types of Tyton® gaskets were used with 4-inch DI pipes in this study: SBR, EPDM, NBR, CR and FKM. Gasolines were purchased from local stores.

Testing Gasket Materials by Equilibrium Sorption

Equilibrium sorption tests were performed on the five types of gaskets in four different grades of gasoline including premium gasoline, regular gasoline, 10% ethanol gasoline (E10) and premium gasoline + 10% MTBE. Specimens for the tests were cut to a thickness of 0.6 cm (0.25 in.) and 1.3 cm (0.5 in.). The samples were immersed in gasoline placed in Teflon®-lined screw-capped jars and kept at room temperature (23 ± 1.5°C). The samples were removed at intervals from hourly to daily, wiped dry with paper towels, and weighed using an electronic balance with an accuracy of 0.001 g. The specimens were returned to the jars immediately after weighing. The samples were weighed until constant weight was reached.

Pipe-drum Apparatus

To simulate a buried pipe in contaminated soil, a pipe-drum apparatus was developed. The drum has a nominal volume of 20-L (5 gallons). The bell and spigot ends of the pipe were cut to lengths of 17.8 cm (7 in.) and 13 cm (5 in.), respectively. The ends of the pipes were welded with steel plates equipped with fittings for connecting ¼ inch copper tubing and valves for sampling. The fittings were sealed with chemically resistant epoxy putty, described in Chapter 3. The bell and spigot were assembled according to the manufacturer's specification. Efforts were made to ensure that the bell and spigot were properly aligned. The pipe-joints were restrained from separating by steel straps bolted together with threaded rods (Figure 6.1). The volume capacities of the pipe-joints were measured before installation in the drum. The pipe-joints were placed in the 20-L drums and the fittings oriented so that the pipe-joints could be filled without trapping air. Sand was then added to the drum.

The copper tubing from the sampling fittings was passed through two holes in the lid and connected to the sampling valves, The holes in the lid were sealed with gasoline-resistant silicone caulking. The lid was secured to the pail with a lever-lock. For apparatuses to be used with aqueous contaminant media, two additional holes were drilled through the drum to accommodate copper tubing used for replenishment of the contaminant solution. Fresh aqueous solution was added by using a copper tube with one end reaching to the bottom of drum and terminating in a loop with holes for discharging the replenishment solution. Overflowing solution was then drained by the other copper tube, which was placed several centimeters below the lid. This design allowed for filling and flushing from bottom to top and minimized volatilization losses during replenishment.

The pipe-joint was filled with deionized water from the bottom through the valve connected to the lower fitting. Samples were taken by gently applying compressed air into the apparatus through the valve connected to the upper fitting. Samples for analysis were collected in 40 mL vials. The vials were capped using Teflon®-coated silicone rubber seals. The samples were analyzed by purge and trap gas chromatography as described previously. After each sampling, the pipe was rinsed and flushed three times with deionized water before the final filling and closing of the valves. Samples were collected weekly or monthly.

Experimental Matrix

Pipe-drum apparatuses for simulating permeation by free product gasoline were set up with the five different types of gaskets and premium gasoline was added to each until it covered

Figure 6.1 Pipe-joint assembly

the sand. Another set of pipe-drum apparatuses were set up using silica sand and water saturated with gasoline (100%) and 50%, 20% and 5% saturated aqueous solutions of gasoline using SBR and NBR gaskets (Figure 6.2). One pipe-drum apparatus used a FKM gasket and water saturated with gasoline. The aqueous solutions of gasoline were replaced weekly to monthly by pumping fresh solutions into the sand with a Masterflex pump. As shown in Table 6.1 and Table 6.2, approximately 68%, 66%, 67% and 79% of the original concentrations remained at the end of one month. Gasoline-saturated water (100%) was prepared by mixing 1 L premium gasoline and 10 L deionized water in a 12-L glass bottle for 48 hours with constant stirring. Lower levels of saturation were prepared by diluting the 100% saturated solution with deionized water.

Hydrostatic Testing

The effect of hydraulic gradient on permeation through SBR gaskets was tested at pressures of 20, 40 and 60 psig. A schematic drawing of the apparatus is shown in Figure 6.3. The heavy duty restraint system shown in Figure 6.4 was necessary to restrain the joint at pressures above 20 psig. The lighter duty straps and rods used at pressures from ambient to 20 psig bent and broke at higher pressures. To pressurize the pipe, the lower valve was closed and air from an air compressor was used to pressurize the apparatus through the valve connected to

Figure 6.2 Pipe-drum apparatuses for aqueous gasoline-saturated solution experiments with SBR, NBR, and FKM gaskets

Table 6.1
Concentrations of aqueous gasoline solutions before introduction into pipe-drum apparatuses

	5% aqueous solution of gasoline (μg/L)	20% aqueous solution of gasoline (μg/L)	50% aqueous solution of gasoline (μg/L)	100% aqueous gasoline-saturated solution (μg/L)
Benzene	3,103 ± 545	11,525 ± 1,683	35,159 ± 3,451	68,444 ± 3,483
Toluene	3,693 ± 1,115	13,248 ± 3,456	42,167 ± 9,865	79,417 ± 5,593
Ethylbenzene	181 ± 75	557 ± 194	1,685 ± 492	3,091 ± 344
m-xylene	544 ± 231	1,525 ± 565	4,479 ± 1,429	8,236 ± 1,072
o+p-xylene	591 ± 248	1,685 ± 627	4,953 ± 1,561	8,985 ± 1,159
Total BTEX	8,111 ± 2,071	28,539 ± 6,160	88,431 ± 16,282	168,172 ± 6,563

Table 6.2
Concentrations of aqueous gasoline solutions after draining from pipe-drum apparatuses

	5% aqueous solution of gasoline (μg/L)	20% aqueous solution of gasoline (μg/L)	50% aqueous solution of gasoline (μg/L)	100% aqueous gasoline-saturated (μg/L)
Benzene	2,443 ± 1,127	7,914 ± 3,272	26,780 ± 5,974	49,055 ± 7,409
Toluene	2,930 ± 1,104	8,846 ± 2,552	26,301 ± 7,500	53,068 ± 3,507
Ethylbenzene	150 ± 87	354 ± 68	802 ± 262	1,774 ± 309
m-xylene	457 ± 289	991 ± 171	2,160 ± 712	4,718 ± 858
o+p-xylene	496 ± 307	1,093 ± 182	2,435 ±794	5,276 ± 1,139
Total BTEX	6,477 ± 2,812	19,199 ± 6,240	58,479 ± 14,519	113,892 ± 3,490

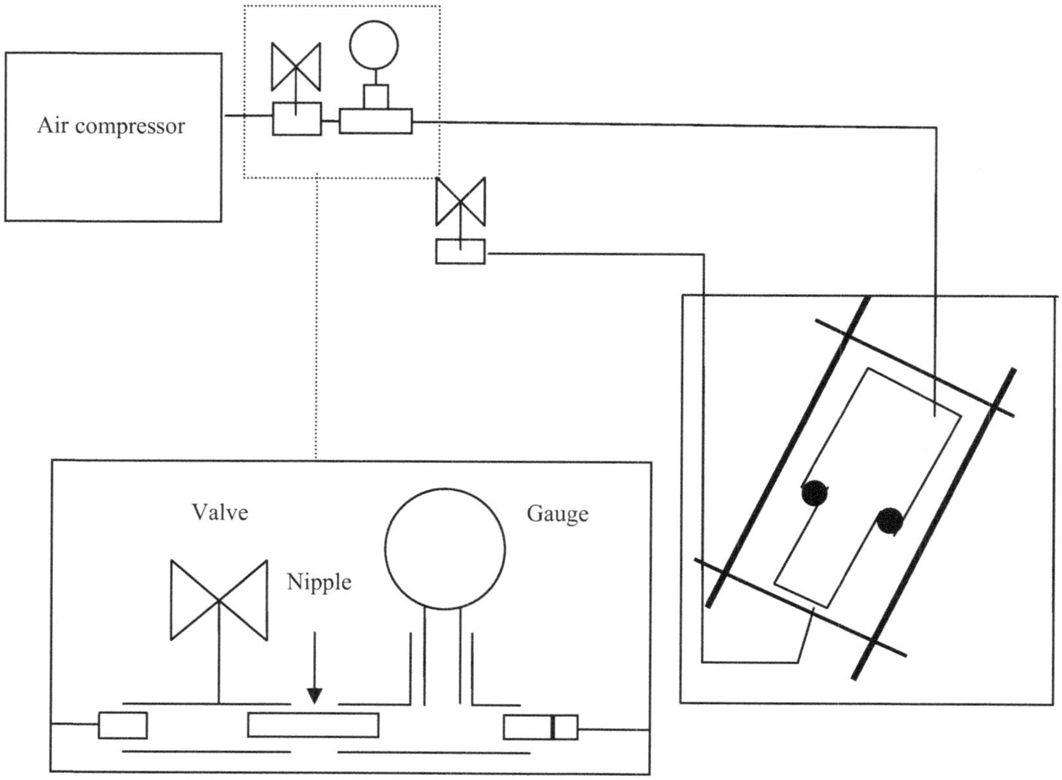

Figure 6.3 Hydrostatic pressure testing scheme

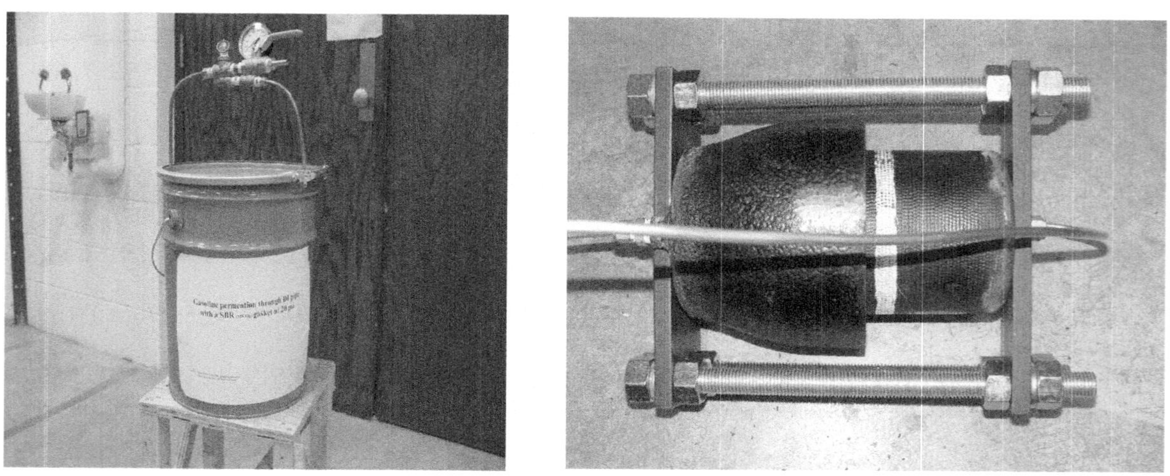

Figure 6.4 Hydrostatic pressure test apparatus (left) and the pipe joint apparatus (right)

the upper fitting. When the desired pressure was reached, the upper valve was closed. By first filling the pipe-joint and the copper tubing with water, air was excluded from the pipe-joint apparatus during pressurization.

Analysis of Data

The cumulative mass of contaminant permeated (mg/joint/day) through the pipe joint was computed at each sampling event from the analytical data and the volume of water in the pipe-joint. When the rate of permeation reached steady state, the permeation rate was estimated by calculating the slope of the plot of the cumulative mass transfer versus time. Permeation rates were computed for total BTEX as well as for the individual BTEX compounds.

RESULTS AND DISCUSSION

Testing Gasket Materials by Equilibrium Sorption

The results of sorption tests for five different Tyton® gasket materials in three different types of gasoline and gasoline + 10% MTBE are presented in Figure 6.5. Except for E10, the

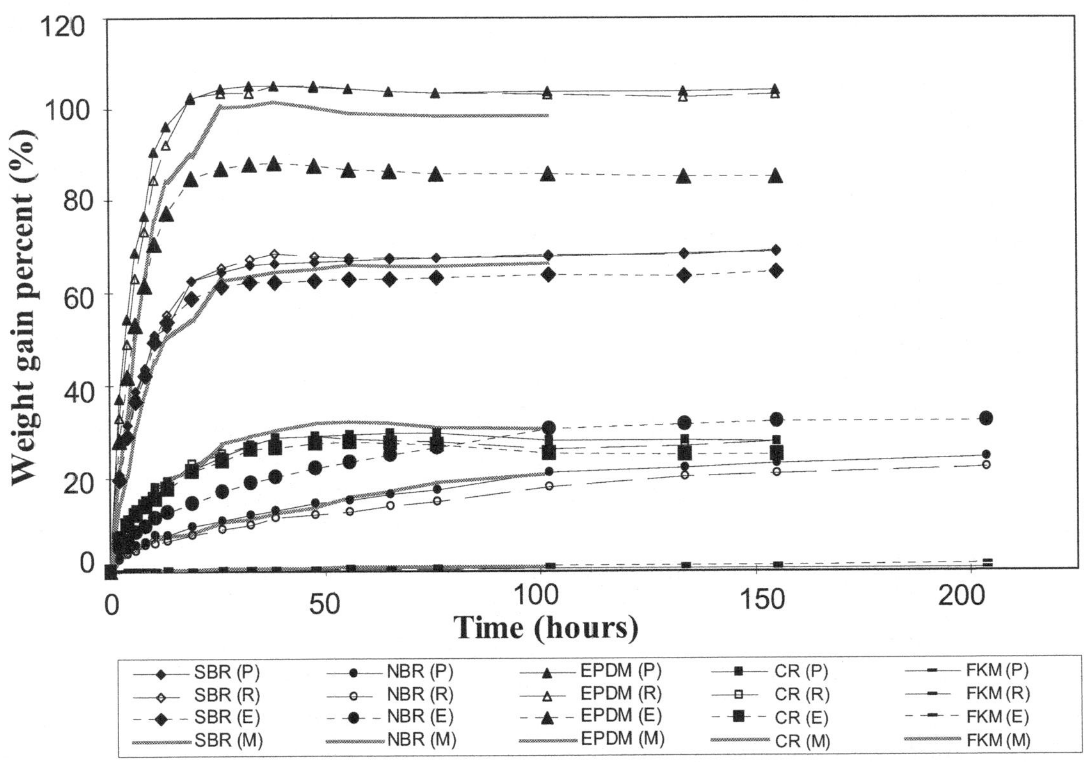

Figure 6.5 Equilibrium sorption tests for 5 different gasket materials in 4 contaminants (P: premium gasoline, R: regular gasoline, E: 10% ethanol gasoline (E10) and M: premium+10% MTBE)

order of percent weight gain for gasoline, from highest to lowest, was EPDM > SBR > CR > NBR > FKM. For E10, the order of weight gain (highest to lowest) was EPDM > SBR > NBR > CR > FKM. Equilibrium sorption for the SBR, NBR and EPDM gaskets was reached within one day. CR and FKM gaskets showed a slow increase with time. Since the sorption of premium gasoline for all gaskets showed higher percent weight gain than other gasoline, premium gasoline was used to conduct the pipe-drum permeation experiments to correlate with the sorption results.

The sorption data plotted as the percent weight gain versus the square root of time, $t^{1/2}$ (min), for premium gasoline are presented in Figure 6.6 and the equilibrium (maximum) percent weight gains for five different gasket materials are presented in Table 6.3. EPDM showed high sorption (127%) in comparison to the other gaskets, while FKM, known for high resistance to gasoline, exhibited very low sorption (about 1%). The percent weight gains for each gasket from highest to lowest were EPDM > SBR > CR > NBR > FKM, which was also reflected by the sizes of the swollen gaskets after exposure to gasoline (see Figure 6.7). The sorption behavior of a polymer is dependent on the characteristics of the polymer structure and the nature of solvent. Uptake of solvent can be explained by solvent molecules moving into the polymer space (lattice) by relaxing the chain segments of polymers (Khinnavar et al. 1992). The sorption results suggest that EPDM chain segments relax more in the presence of gasoline than the other polymers tested.

Tyton® gasket indicates are "dual durometer" gasket consisting of a hard "heel" and a soft "bulb." Figure 6.8 shows a cross-section of a 4-inch SBR Tyton® gasket. The heel anchors the gasket in place during assembly of the joint and the bulb forms the hydrostatic seal. The heel and bulb have different resilience characteristics which suggest different polymer formulations and possibly different permeation characteristics. The lengths and widths of the cross-section (see Figure 6.8) of the SBR gaskets were measured and were found to be 1.00 ± 0.03 ins. and 0.60 ± 0.02 ins, respectively. The dimensions of the five different types of gaskets were measured and were found to identical.

Sorption of premium gasoline by the heel and bulb rubbers of the five different gaskets was tested. The sorption curves of the heel and bulb portions of the five gasket specimens are as shown in Figure 6.9 and were found to increase almost linearly with the square-root-of-time in the initial stage before reaching equilibrium sorption. This initial linear portion of the sorption curve generally implies Fickian diffusion during this period. In Fickian diffusion, the polymer segments swollen by the solvent molecules and are rearranged to sorb more solvents (Harogoppad and Aminabhavi 1992).

Comparison of percent weight gains by the heel and bulb portions showed that the bulb sorbed more than the heel for all gasket materials except for CR (see Table 6.3) and there were only slight differences in swelling between heel and bulb rubbers of NBR and FKM. The difference in sorption between the heel and bulb of the SBR gasket was particularly large with the bulb sorbing about 60% more premium gasoline than the heel for a weight gain of 97% over its original weight. Although the heel and bulb rubber consist of the same basic types of rubber, the heel is compounded with extra carbon black to make it harder (Bird 2006). Gaskets with a high swelling tendency reached equilibrium sorption faster than gaskets that did not swell much.

To identify where the gasoline came in contact with the surfaces of the gasket in a pipe joint, paint was applied to a gasketed pipe joint (Figure 6.10). The gasket was disassembled after one week. The white paint on the gasket as shown in Figure 6.10 indicated that the heel of the gasket was most likely to be exposed to the contaminants. This area probably represents the contact surface of a gasket with outside organic compounds after the pipe is jointed and the gasket is compressed.

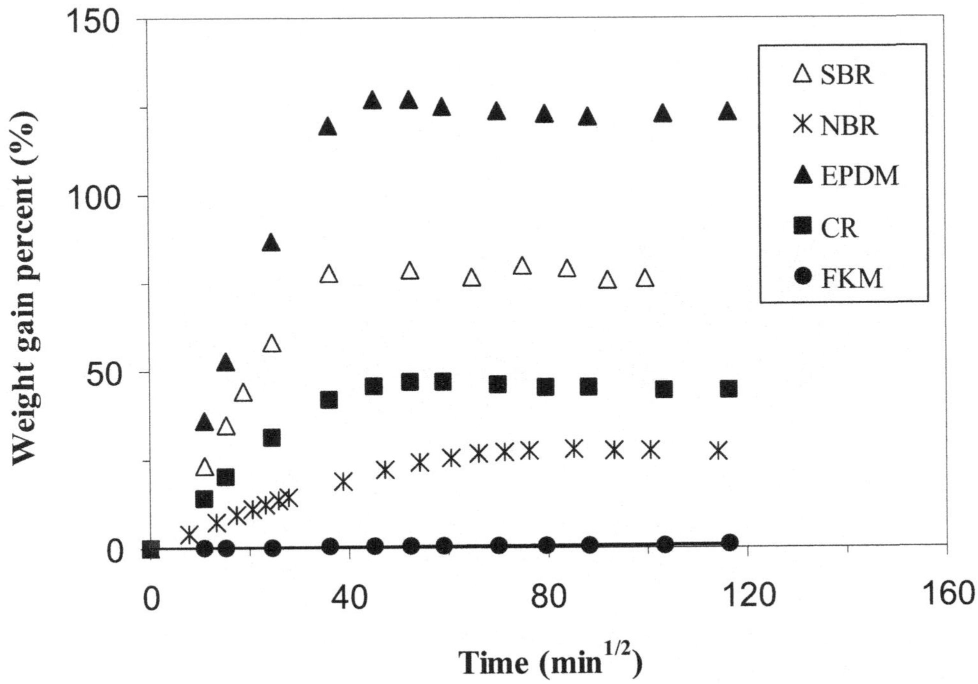

Figure 6.6 Sorption tests for 5 different gasket materials with premium gasoline

Table 6.3
Percent weight gain of gasket materials in premium gasoline

	Total (Heel + Bulb)	Heel	Bulb
SBR	80.0 %	61.1 %	97.3 %
NBR	27.7 %	26.9 %	29.0 %
EPDM	127.0 %	97.3 %	141.6 %
FKM	0.65 %	0.81%	0.82 %
CR	46.9 %	57.1 %	43.8 %

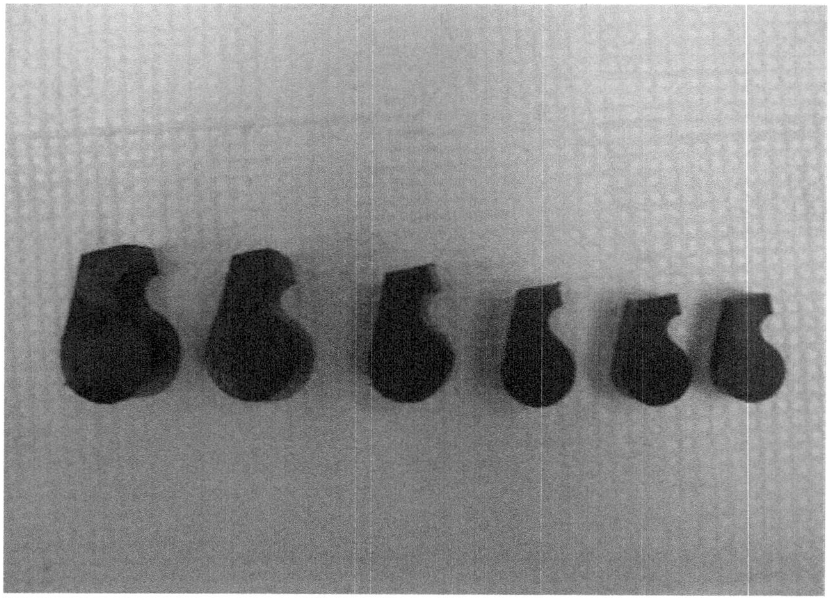

Figure 6.7 Gaskets exposed to gasoline, from left to right - EPDM, SBR, CR, NBR, FKM, and an unexposed SBR gasket.

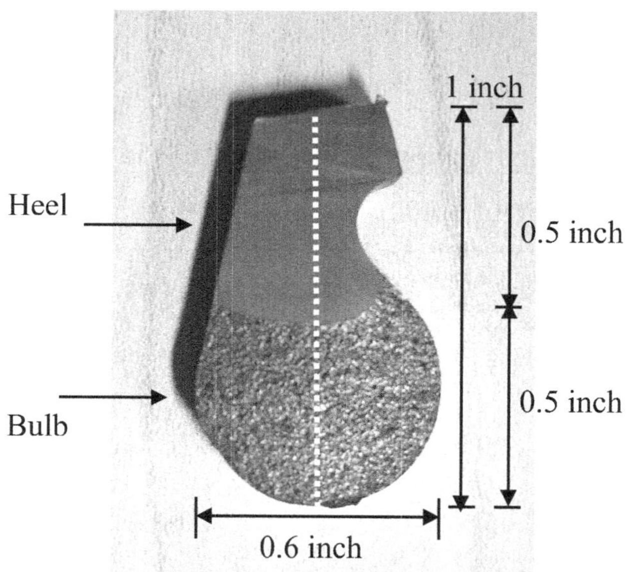

Figure 6.8 Cross section and dimensions of a SBR gasket showing heel and bulb portion

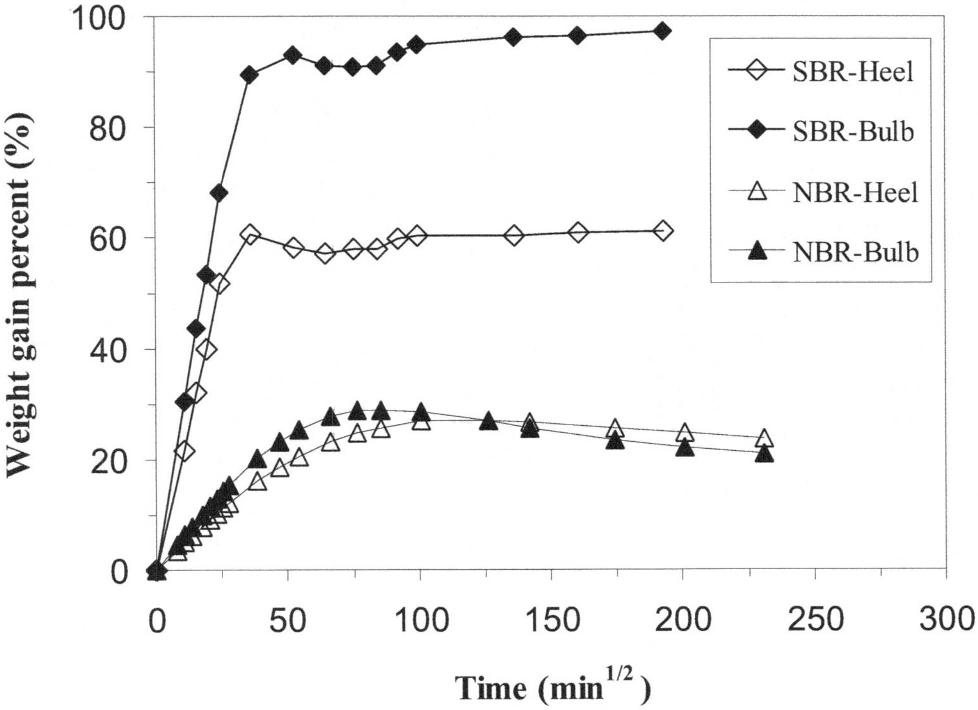

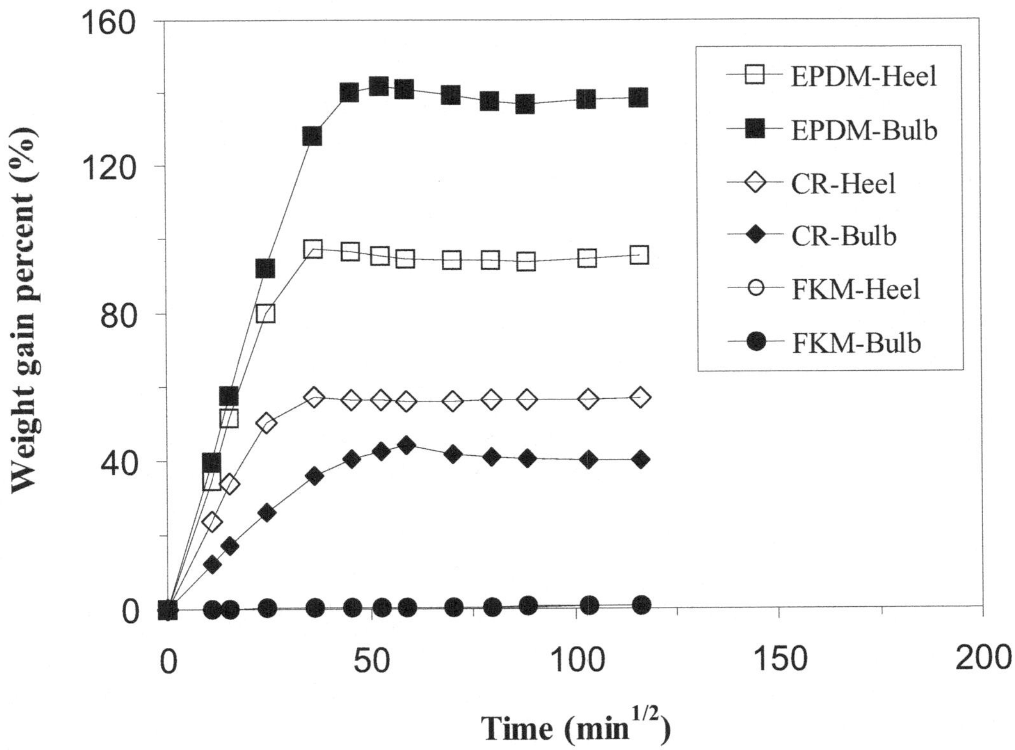

Figure 6.9 Sorption tests for heel and bulb of five gasket materials with premium gasoline

Figure 6.10 Measurement of possible gasoline contact area using paint

Premium Gasoline Permeation Experiments

SBR, NBR, CR, EPDM, and FKM gaskets were tested in Pipe-drum apparatuses using premium gasoline. The permeations of the gaskets by BTEX are shown in Figure 6.11 and the permeation rates summarized in Table 6.4.

Breakthrough for the EPDM gasket was first detected after 35 days of exposure. This was followed by CR and SBR gaskets at 50 days. After 124 days of exposure, BTEX breakthrough for the NBR gasket was detected. NBR gasket exhibited more resistance to gasoline permeation than SBR, EPDM and CR gaskets. This is in agreement with the equilibrium sorption experiments in which EPDM had the highest weight gain percent compared to other gaskets.

Low concentrations of BTEX compounds were detected within 14 days for the FKM gasket and persisted throughout the experiment. These concentrations were fairly constant indicating that there might been a leak rather than permeation. FKM is a very hard material compared to the other gaskets. Joining the bell and spigot with the FKM gasket required considerable effort since the hardness of the FKM gasket made it difficult for the pipe segments to slip together. This hardness is also reflected in the very low sorption of the FKM material (less than 1% weight gain).

The total BTEX permeation rates for the SBR, CR and EPDM gaskets were approximately 4 mg/joint/day while the total BTEX permeation rates for NBR and FKM gaskets were 0.3-0.5 mg/joint/day (Table 6.4). The order of breakthrough for BTEX compounds for EPDM, CR, SBR and NBR gaskets was benzene > toluene > ethylbenzene > m-xylene > o+p-xylene. Benzene and toluene are the main BTEX compounds in gasoline and have higher solubility in water than the other BTEX compounds. The permeation rates for each individual BTEX compounds were estimated and are presented in Table 6.4.

Table 6.4
Permeation rates (mg/joint/day) through 4-inch Tyton® gaskets

Gasket	Benzene	Toluene	Ethylbenzene	m-Xylene	o+p Xylene	Total BTEX
EPDM	1.06	2.25	0.12	0.29	0.31	3.93
SBR#	2.72	2.22	0.068	0.164	0.185	5.20
CR	1.47	2.34	0.08	0.19	0.22	4.23
NBR	0.16	0.12	0.004	0.008	0.008	0.36
FKM*	0.15	0.23	0.013	0.030	0.033	0.49

[#: average values from 8 experiment data; *: possible leakage]

Aqueous Gasoline Solution Experiments

Permeation of BTEX through SBR in 100% and 50% saturated aqueous solutions of gasoline was observed after 7 months of exposure as shown in Figure 6.11. The breakthrough time for SBR exposed to gasoline-saturated water (100%) was 4 times longer than that for pure premium gasoline. The total BTEX permeation rate was estimated to be 0.203 mg/joint/day in gasoline-saturated water while the permeation rate was 5.20 mg/joint/day in the pure premium gasoline. Total BTEX concentrations in premium gasoline and aqueous gasoline-saturated solution (100%) were approximately 172,000 mg/L and 168 mg/L, respectively. BTEX concentrations in premium gasoline and 100% gasoline-saturated solutions appeared to impact the permeation rate and breakthrough time. No permeation by BTEX compounds through SBR gaskets exposed to 20% and 5% saturated aqueous solutions of gasoline was observed after 18 months of exposure. It should be noted that the BTEX concentration in a 20% saturated aqueous solution of gasoline is approximately 35 mg/L, which is typically higher than BTEX concentrations found in a contaminated plume in the field.

There was no breakthrough of BTEX compounds after 550 days (18 months) of exposure to gasoline-saturated water (100%) for NBR and FKM gaskets (Figure 6.11 and Figure 6.12). There were also no breakthrough of BTEX for NBR gaskets exposed to 50%, 20%, and 5% saturated aqueous solutions of gasoline.

Hydrostatic Testing

Hydrostatic testing results indicated that permeation of BTEX compounds through SBR gaskets exposed to premium gasoline was not impacted by pressure as shown in Figure 6.13. The first set of experiments showed that the permeation rate of SBR gaskets at 0 psig was higher

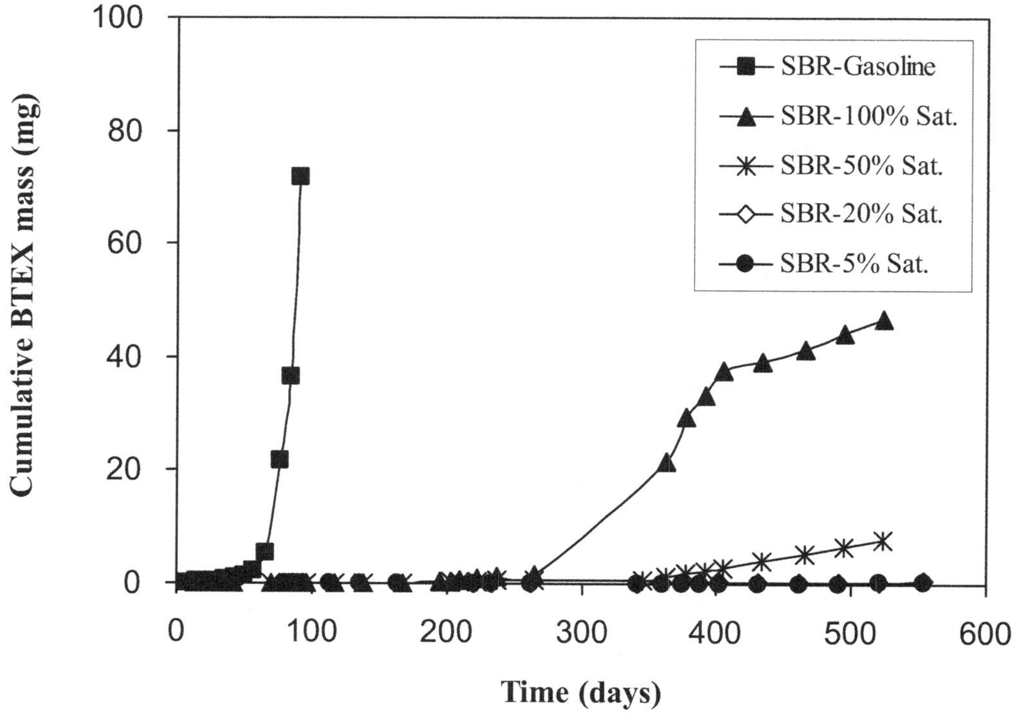

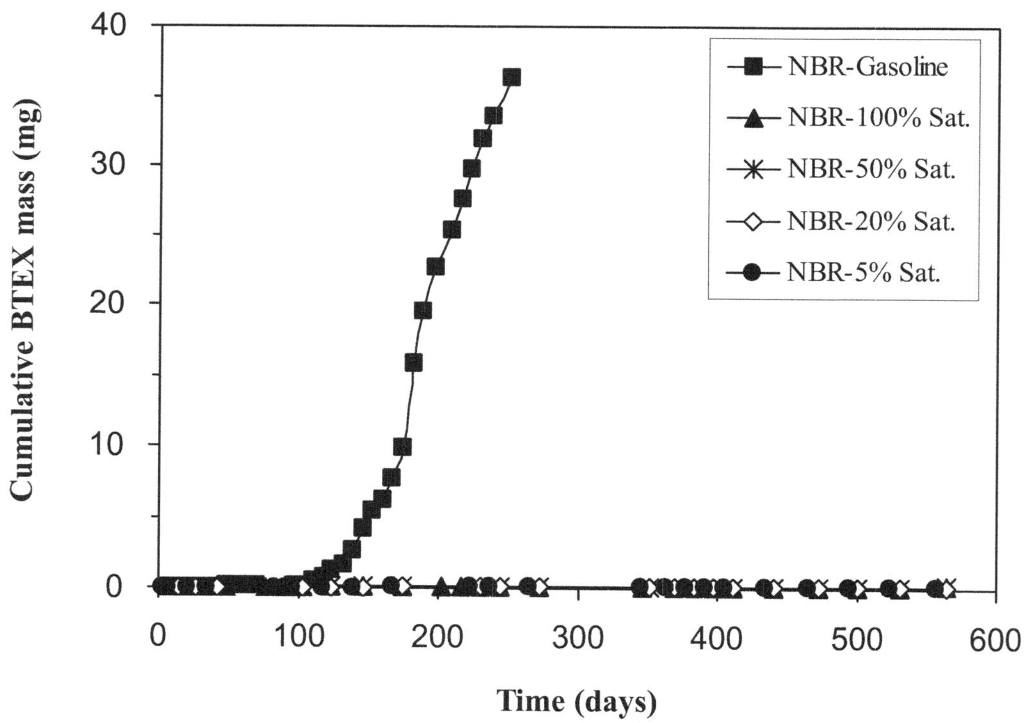

Figure 6.11 Permeation of total BTEX through SBR and NBR gaskets

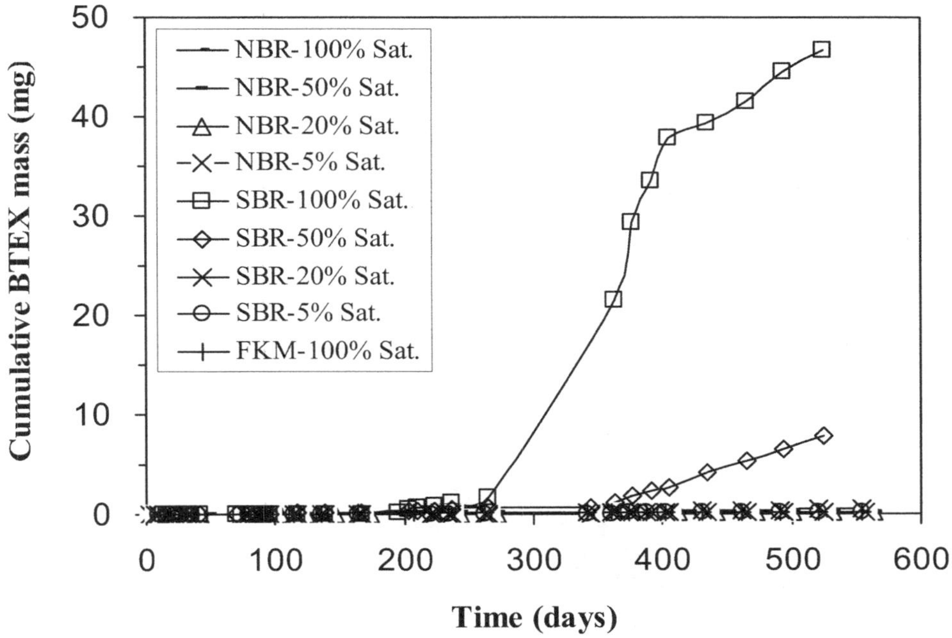

Figure 6.12 Permeation of total BTEX through SBR, NBR and FKM gaskets in gasoline-saturated solutions (100%), 50%, 20%, and 10% saturated aqueous solutions of gasoline

than at 60 psig but lower than at 20 and 40 psig. Based on these results, a second set of experiments was performed to verify that hydrostatic pressure had no effect on gasoline permeation. For this second set of experiments, the permeation rate at 0 psig was found to be significantly lower than the other pressures and the elevated pressures did not correlate with those of the first set of experiments.

The permeation rates of SBR Tyton® gaskets in DI pipes for four different levels of hydrostatic pressure tests were plotted in Figure 6.14. One experiment at 0 psig total BTEX (18.1 mg/joint/day) was considered to be an outlier with a 95% probability based on calculations using Dixon and Grubs tests and was not plotted on Figure 6.14. From Figure 6.14, the scatter indicates that there was no correlation between the hydrostatic pressures and the permeation rates (R^2=0.129).

Two possibilities may explain this discrepancy: (1) differences in the composition of gaskets or construction; and (2) misalignment. The Tyton® gasket is made of two parts - the "heel" and "bulb". The materials for the heel and the bulb have different sorption capacities. The bulb was found to sorb about 60% more gasoline than the heel. While the bulb provides the hydrostatic seal in the water main, both the heel and bulb are part of the permeation pathway for external contaminants. The earlier experiment with paint indicates that exterior contact with contaminated solution is mainly with the heel. Little or no bulb rubber is exposed directly to external solution. Permeation must proceed first through the heel rubber and then through the more permeable bulb rubber. Since the heel is the rate limiting material, it is possible, that due to differences in exposure area of the heel, the path length taken by the contaminants may vary from gasket to gasket.

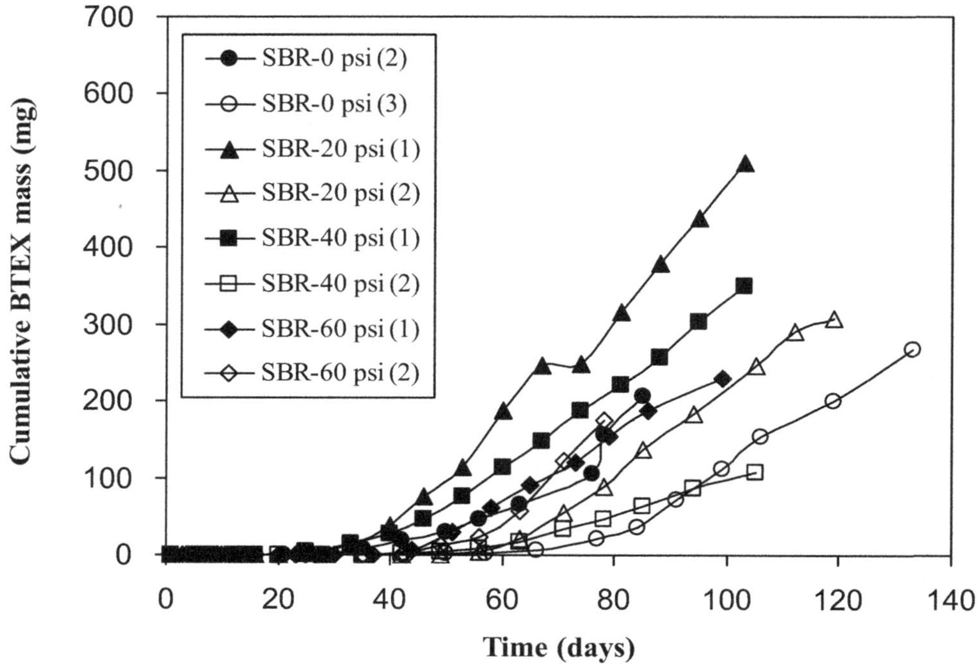

Figure 6.13 Permeation of total BTEX through SBR gaskets at 0, 20, 40 and 60 psig ((1) – first round of tests, (2) - second round of tests, (3) repeat of first round outlier)

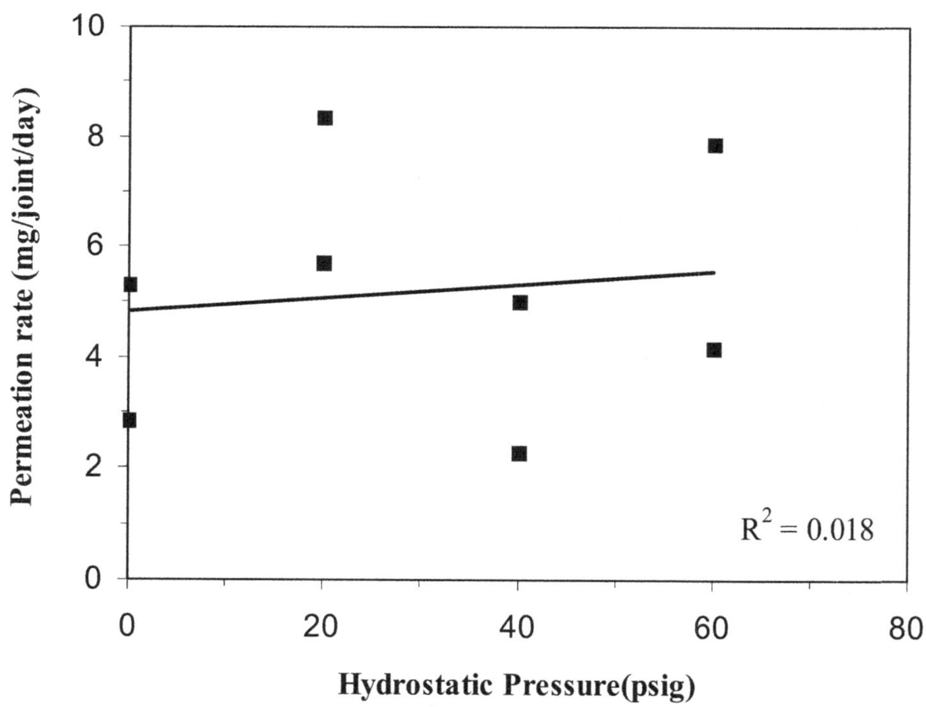

Figure 6.14 Permeation of total BTEX through SBR gaskets versus hydrostatic pressure

Glaza and Park (1992) indicated that misalignment may have an impact on permeation through the gasket. They tested three SBR gasket pipes which were misaligned by about 3°. One of three had a lower permeation rate than other two SBR gaskets. As the maximum deflection angle recommended by the manufacturers is typically in a range of 3° - 5° for DI pipes, this discrepancy may happen under field conditions. However, care was taken in these experiments to ensure that the pipe-joints were properly aligned when installed.

Correlation of Sorption Data and Permeation Rates

Since the heel is in contact with the external contaminants and the sorption of the heel is much less than that of the bulb, the heel would determine the overall permeability of SBR Tyton® gaskets. Table 6.5 provides the estimated total BTEX permeation rates for 4-inch Tyton® gaskets, the equilibrium sorption (percent weight gain), and sorption rate (percent weight gain/min$^{1/2}$) for the heel rubbers. The EPDM gasket had a 98% weight gain from the equilibrium sorption test which was also reflected in the high permeation rate. Similarly, both CR and SBR gaskets had more than 50% weight gains but had permeation rates similar to that of EPDM. The percent weight gain for NBR was about 27% but the BTEX permeation rate of 0.36 mg/joint/day was very low. In general higher sorption rates were observed for gaskets that had high permeation rates.

Based on the above observations, simple linear correlations between the permeation rates and the percent weight gains or sorption rates were attempted as shown in Figure 6.15. The R-squared value for a linear regression for permeation rates and sorption rates was 0.73, while the R-squared value was 0.63 for permeation rates and sorption weight gains. Due to the small number of samples in this data set, these correlations should be used with caution to predict the permeation rates.

Table 6.5
Heel rubber sorption and gasket permeation rates for premium gasoline

Experiments	4-inch Tyton® Gaskets				
	EPDM	CR	SBR	NBR	FKM
Equilibrium sorption for Heel (% wt gain)	97.3%	57.1%	61.1%	26.9%	0.81%
Sorption rate for Heel (% wt gain /min$^{1/2}$)	3.27	2.06	2.09	0.38	0.0073
Permeation rate (mg BTEX/day)	3.93	4.23	5.20$^{\#}$	0.36	0.49*

[$^{\#}$ average of 8 values; * possible leakage]

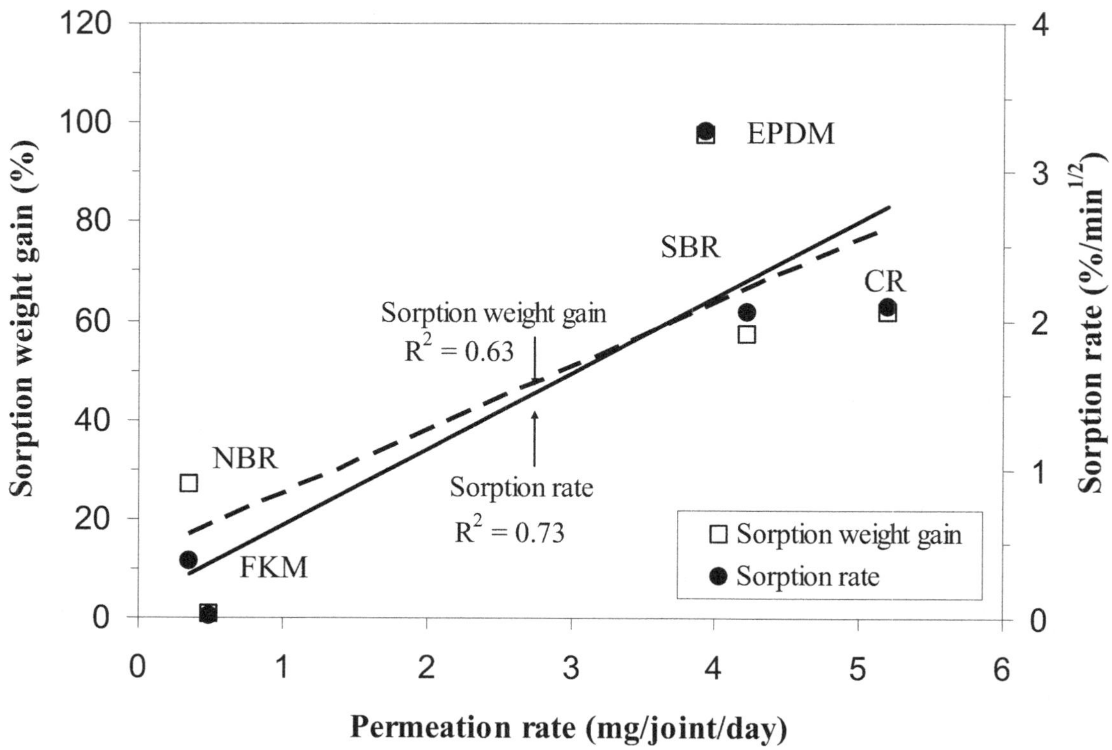

Figure 6.15 Correlation of results from equilibrium sorption of heel portion and permeation data from pipe-drum experiments

Scale-up of Permeation in DI Pipe

The mass permeated per joint per day, M, and the permeation rate, P_m, in a pipe joint are related by following equation.

$$P_m = \frac{M}{A} \tag{6.1}$$

where P_m = permeation rate in one pipe joint ($\mu g/cm^2/day$)
M = mass permeated per joint per day (μg/joint/day)
A = area of the gasket contacted by solvents (cm^2)

Using Fick's First Law

$$P_m = -D\frac{\partial C}{\partial x} = -D\frac{C_o - C_i}{\Delta x} \tag{6.2}$$

where D = diffusion coefficient of gasket material (cm^2/s)
C_o = solution concentration inside the pipe joint (μg/L)
C_i = solution concentration outside the pipe joint (μg/L)
Δx = permeation distance (cm)

By assuming the inside concentration, C_i, to be zero (drinking water within the pipe) and the thickness (i.e., length of permeation path) of the gasket to be L, the equation becomes,

$$P_m = -D\frac{C_0 - 0}{L} \tag{6.3}$$

For same type of gasket material (i.e., the same diffusion coefficient) and the same external concentration, the equation is given by:

$$P_{m,1}L_1 = P_{m,2}L_2 = -D(C_0 - 0) \tag{6.4}$$

Substituting for the permeation rate, P, with equation 6.1, the equation becomes,

$$\frac{M_1}{A_1}L_1 = \frac{M_2}{A_2}L_2 \tag{6.5}$$

Rearranging:

$$M_2 = M_1\frac{L_1}{L_2}\frac{A_2}{A_1}, \text{ or } M_2 = M_1\frac{\beta_A}{\beta_L} \tag{6.6}$$

where β_L = length factor ($= L_2/L_1$)
β_A = area factor ($= A_2/A_1$)

Knowing the dimensions of two gaskets, equation 6.6 can be used to estimate the permeation rate for pipes of any diameter using the experimentally determined permeation rate of a single pipe of known diameter.

Permeation of Flowing Mains

Engineering decisions regarding permeation of gaskets by gasoline should be based on the rate of permeation by benzene since the MCL for benzene (5 µg/L) is 200 times lower than that of toluene (1,000 µg/L) and since benzene is the most soluble in water of the BTEX compounds. In the experiments involving 4-inch SBR Tyton® gaskets exposed to free product gasoline (a worst case scenario), the average rate of permeation by benzene was 2.72 ± 1.28 (standard deviation) mg/joint/day (based on all the experiments conducted with SBR gaskets, including the hydrostatic pressure experiments). Allowing for a safety margin of three standard deviations (99.7% confidence), a benzene permeation rate of 6.56 mg/joint/day may be used to predict benzene concentrations in flowing mains. Assuming this permeation rate, a volume of 347 gallons (1,312 L)/joint/day of water would be required to dilute the concentration of benzene to the MCL of 5 µg/L. This corresponds to an average flow rate of 0.24 gpm/joint in 4-inch DI main. Since the analytical method detection limit (MDL) for benzene (0.24 µg/L for the analytical method used in this study) was about 21 times lower than the MCL, a flow of 5 gpm/joint would dilute benzene permeating the SBR gasket in this worst case scenario to levels that are not analytically detectable.

For the experiment involving a 4-inch NBR gasket exposed to free product gasoline, the benzene permeation rate was experimentally found to be 0.159 mg/joint/day, which means that a flow of 0.0058 gpm/joint would be required to dilute the pollutant concentration to the MCL and 0.12 gpm to be rendered undetectable for one DI pipe joint.

For the 4-inch SBR gasket exposed to 100% saturated aqueous gasoline solution (containing about 168 mg/L total BTEX), the benzene permeation rate was found experimentally to be 0.073 mg/joint/day, which would require 2.7 x 10^{-3} gpm to dilute to the MCL and 0.056 gpm to be rendered undetectable.

Clearly, only a minimal flow in a water main, even in the worst case scenario, will reduce benzene concentrations below the MCL and most likely render benzene undetectable. Since anecdotal evidence and the threshold odor limit of 190 μg/L suggests that the benzene MCL is exceeded before the odor threshold is exceeded, minimal flow should also eliminate taste and odor complaints resulting from permeation of gasoline. According to the Ductile Iron Pipe Research Association (DIPRA), the common flow velocity in DI pipe with cement/cement-mortar linings used for potable water is 2 to 10 feet per second (ft/s) or approximately 80 to 400 gpm for a 4-inch pipe. In realistic conditions, contaminants would be diluted to undetectable levels with a minimum flow of 2 ft/s in a 4-inch pipe.

The permeation rates for benzene in DI pipes larger than 4-inch were estimated using equation 6.6, as presented in Table 6.6. The estimated permeation rates were based on the experimental permeation rates for benzene in 4-inch DI pipe from pipe-drum experiments, which were 6.56 mg/joint/day for SBR gaskets exposed to free product gasoline; 0.159 mg/joint/day for SBR gaskets exposed to gasoline saturated water; and 0.073 mg/joint/day for NBR gaskets exposed free product gasoline. The permeation rate of 6.56 mg/joint/day for SBR exposed to free product gasoline includes a three-standard-deviation margin for a 99.7% confidence level (n=8). The other two permeation rates were from single experiments. The gasket dimensions as presented in Table 6.6 were obtained from the gasket manufacturer. The permeation rate for a single gasketed joint increases with larger sizes of DI pipe since β_A for a larger DI pipe is larger than β_L. For example, the permeation rates for 10, 16, and 24-inch pipe were estimated to be 11.5, 21.0, and 30.7 mg/joint/day, respectively, for SBR gasket exposed to free product gasoline.

As stated earlier, the typical flow velocities in water distribution pipes ranged from 2 to 10 feet per second (ft/s) (0.6 to 3 m/s). Using 2 ft/s, the flow rate for a 24-inch pipe is estimated to be 11,520 gpm. As presented in Figure 6.16, the estimated flow rate needed to dilute the contaminant level to the benzene MCL for 100 feet (5 joints) of a 24-inch DI pipe with SBR gaskets exposed to free product gasoline is approximately 10 gpm. This means that a minimal flow will easily provide sufficient dilution so that the benzene MCL will not be exceeded. In the case of NBR gaskets exposed to gasoline, the flow rate needed to dilute benzene to less than the MCL in a 24-inch pipe is 0.1 gpm.

Permeation During Conditions of Stagnation

The concentrations of benzene that might result from periods of stagnation were estimated using the benzene permeation rates in Table 6.6 and presented in Figure 6.17. Assuming a scenario of 100 ft of 4-inch DI pipe (with 5 gaskets) and containing 247 L of water exposed to free product gasoline, the benzene concentrations after 8 hours of stagnation were estimated to be 19 μg/L for SBR gaskets and 1.1 μg/L for NBR gaskets. In this scenario, the 5 μg/L MCL for benzene was exceeded for SBR gaskets whereas permeation through the NBR

Table 6.6
Estimated permeation of benzene in 4-inch to 24-inch DI pipes with Tyton® gaskets

Pipe Diameter (ins)	Gasket width* (cm)	Contact area** (cm^2)	β_L	β_A	Permeation rate (mg/joint/day)		
					SBR-Gasoline+	SBR-100%++	NBR-Gasoline+
4	3.63	44.41	1.00	1.00	6.56ς	0.073ς	0.159ς
6	3.63	62.28	1.00	1.40	9.20	0.102	0.223
8	4.72	89.63	1.30	2.02	10.18	0.113	0.247
10	5.08	108.94	1.40	2.45	11.51	0.128	0.279
12	5.08	128.72	1.40	2.90	13.59	0.151	0.329
14	5.84	201.63	1.61	4.54	18.52	0.206	0.449
16	5.84	228.22	1.61	5.14	20.96	0.233	0.508
18	5.84	254.81	1.61	5.74	23.40	0.260	0.567
20	5.84	281.40	1.61	6.34	25.84	0.288	0.626
24	5.84	334.57	1.61	7.53	30.73	0.342	0.745

* width (thickness) of gasket as provided by manufacturer
** largest cross section area of gasket (estimated from manufacturer's data)
+ gasket in contact with pure premium gasoline
++ gasket in contact with 100% aqueous gasoline-saturated solution
ς from experimental data

gaskets was low and did not exceed the benzene MCL after 8 hours of stagnation. The estimated benzene concentrations for various pipe diameters using the estimated permeation rates as presented in Table 6.6 indicate that the benzene MCL would be exceeded even for a 24-inch pipe with SBR gaskets after 8 hours of water stagnation for pipes exposed to free product gasoline (see Figure 6.17). However, for DI pipes with SBR gaskets exposed to gasoline saturated groundwater, the benzene concentrations in the pipe would be below the MCL after 8 hours of water stagnation (see Figure 6.17).

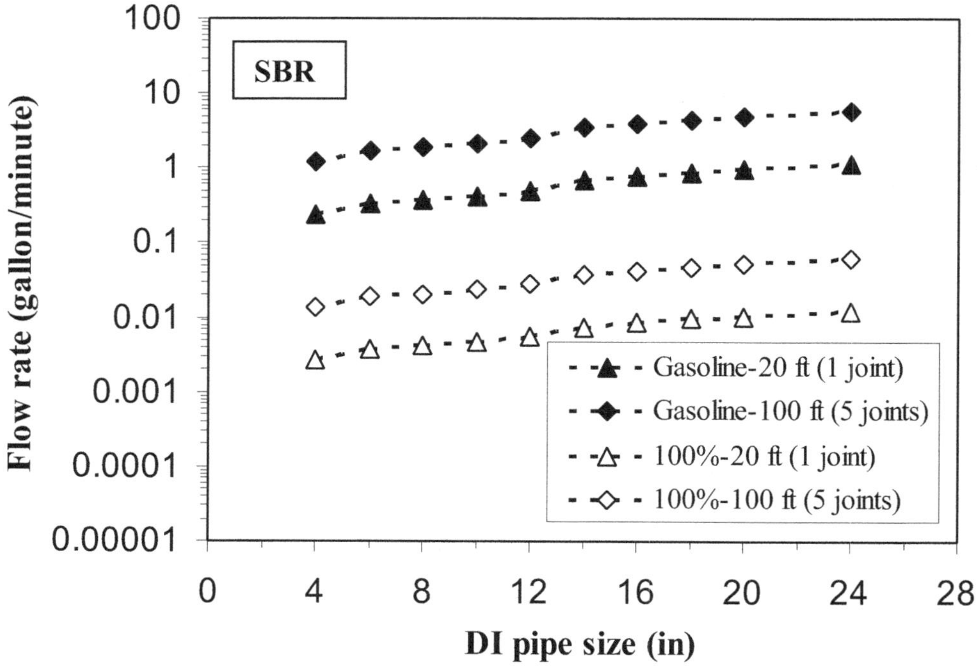

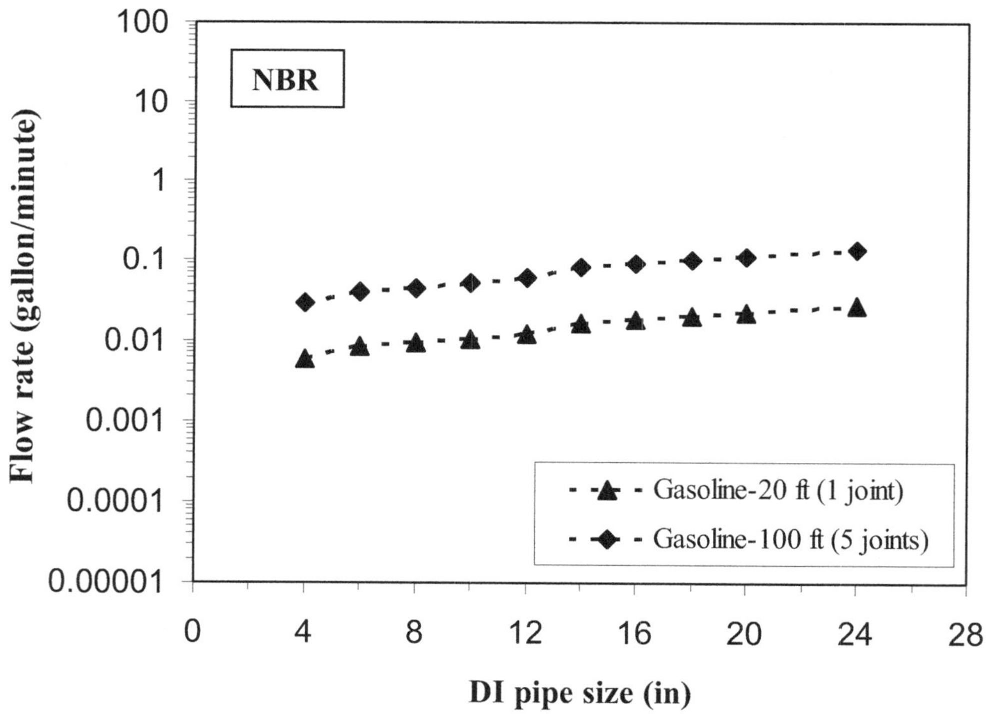

Figure 6.16 Flow rates needed to dilute concentration of benzene to MCL in 4-inch to 24-inch DI pipes with Tyton® gaskets.

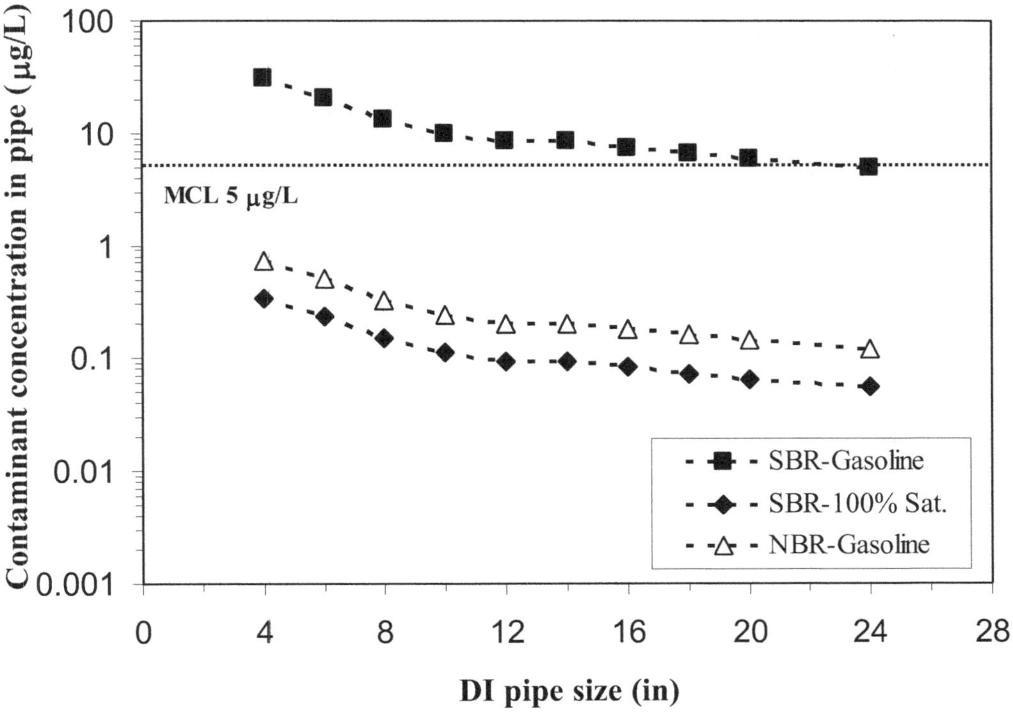

Figure 6.17 Benzene concentrations after 8 hours of stagnation in various sizes of DI pipes with SBR and NBR gaskets (100 ft of pipe with 5 joints)

Note: Concentrations for SBR-gasoline were calculated using a permeation rate equal to the mean obtained from 8 experiments plus 3 standard deviations (99.7% confidence).

CONCLUSIONS

Equilibrium Sorption

Equilibrium sorption experiments showed that bulb and heel portions of certain Tyton® gaskets have significantly different gasoline sorption behavior. The bulb rubber of SBR and EPDM gaskets sorbed significantly more gasoline than the heel rubber, while the bulb and heel of NBR gaskets sorbed nearly equally. The heel rubber of CR gaskets sorbed somewhat more gasoline than the bulb rubber. Sorption of gasoline was very low for FKM gaskets and was not significantly different between bulb and heel. The sorption characteristics of the heel and bulb rubbers probably determine the overall permeability of Tyton® gaskets. Efforts should be make to better understand the permeation pathway and the impact of the bulb and heel rubber on permeation.

Environmental Simulation Using Pipe Drum Apparatuses

Environmental simulation in pipe-drum apparatuses showed that NBR and FKM gaskets are more resistant than other types of gaskets when exposed to premium gasoline. The order of breakthrough of BTEX compounds through Tyton® gasket materials was EPDM > CR > SBR > NBR > FKM. This sequence differs from that of the equilibrium sorption experiments where the order was EPDM > SBR > CR > NBR > FKM.

After more than 550 days of exposure to gasoline-saturated water (100%), no permeation was observed through NBR or FKM gaskets, or through NBR gaskets exposed to lower levels of gasoline contamination. Permeation occurred through SBR gaskets from the 100% and 50% saturated aqueous solutions of gasoline, but not from the 20% and 5% saturated aqueous solutions. Calculations showed that minimal flow in water mains would dilute benzene concentrations in drinking water to below the 5 µg/L MCL, even when SBR gaskets are exposed to free product gasoline. Gasoline in groundwater even at saturated levels will not permeate SBR gaskets at a rate high enough to exceed the benzene MCL. NBR gaskets will protect from exceeding the benzene MCL after 8 hours of stagnation, even when in contact with free product gasoline. SBR Tyton® gaskets are satisfactory for use in any condition of gasoline contamination unless stagnation is expected, in which case NBR is a sufficiently resistant alternative material.

Hydrostatic Pressure

Based on the experimental tests, hydrostatic pressure within a water main does not influence the permeation of hydrocarbons through SBR gaskets.

Scaling-up

Estimated permeation rates for larger sizes of gasketed DI pipes indicated that the benzene MCL of 5 µg/L would be exceeded for SBR gaskets of all sizes (up to 24-inch) when exposed to free product gasoline with 8 hours of water stagnation. However, the permeation rate through NBR gaskets is sufficiently low that the pipe-water would not exceed the benzene MCL for the same conditions. For SBR or NBR gaskets exposed to any level of gasoline contamination in groundwater, even 100% aqueous saturation, the benzene concentration in the pipe-water would not exceed the MCL, even after 24 hours of stagnation.

CHAPTER 7
GASKETED POLYVINYL CHLORIDE (PVC) PIPES - LABORATORY STUDIES UNDER SIMULATED ENVIRONMENTAL CONDITIONS

OVERVIEW

Buried PVC pipelines are commonly joined through the use of an elastomeric gasket. The most widely used method for joining PVC pipes is the Rieber joint system in which the gasket is integrated into the pipe bell during the manufacturing and belling process. This Rieber joint system has been fully adopted by the PVC pipe industry in the U.S. and Canada as it offers many advantages such as high resistance to infiltration and exfiltration, withstanding high internal pressure and vacuum, and preventing leakage when axial joint deflection takes place within allowable limits (Rahman and Alchin 2005).

Of importance is the need to know the ability of the Rieber joint system to resist permeation of organic compounds under typical field conditions. Experimental results in Chapter 4 indicated that PVC pipe material itself is an effective barrier against long-term exposure to commercial gasoline. However, the elastomeric gasket materials such as styrene-butadiene rubber (SBR) and nitrile rubber (NBR) are quite susceptible to permeation by organic contaminants (Glaza and Park 1992). The experiments using gasketed ductile iron pipes (Chapter 6) also demonstrated that gasoline can permeate through the gaskets of the pipe joints rapidly. From the perspective of resistance to permeation, the gasket is the weaker link in gasketed PVC pipes.

The objective of this study was to investigate the impact of petroleum-based hydrocarbons (BTEX) on gasketed PVC pipes equipped with the Reiber gasket system under simulated field conditions, including subsurface gasoline spills and gasoline-contaminated groundwater with various levels of contamination. The results obtained will provide guidance to utilities regarding conditions under which the jointed PVC pipes can be used in gasoline-contaminated areas.

MATERIALS AND METHODS

Materials

Four 2-inch pipe systems were tested. The pipes were: (i) gasketed PVC pipes (blue colored) with styrene-butadiene rubber (SBR) gaskets; (ii) ungasketed PVC pipes (blue colored); (iii) gasketed PVC pipes (white colored) with nitrile (NBR) gaskets; and (iv) ungasketed PVC pipes (white colored). Pipes of the same color were obtained from the same pipe manufacturer. Two-inch pipes were used as they allow for easy installation and manipulation in the experimental apparatus.

Premium gasoline was purchased from a local store and the concentrations of benzene, toluene, ethylbenzene, m-xylene, and o+p-xylene were determined using gas chromatography to be 26.7 g/L, 106.4 g/L, 13.7 g/L, 39.4 g/L, and 38.4 g/L, respectively. BTEX compounds were roughly about 28% by volume of the gasoline. Silica sand (Granusil 4030) was purchased from

UNIMIN Corporation, Portage WI. The sand contained very low organic matter and had a uniform particle size diameter of 0.25 mm.

Experimental Apparatus

To simulate a buried pipe in contaminated soils, a pipe-drum apparatus was developed as shown in Figure 7.1. The apparatus consists of a 20-L drum with the gasketed PVC pipe mounted vertically through holes drilled in the lid and bottom of drum. The ends of the pipes were capped with PVC caps and equipped with brass needle valve assemblies for filling and draining of the pipes. Pipes were filled from top to bottom using a Masterflex pump and drained by gently applying compressed air from the top. Two holes were drilled through the drum lid and two copper tubes (influent and effluent) installed to allow aqueous gasoline-saturated solutions to be pumped with a Masterflex pump into the drum to renew the solution in the drum. All connections were sealed with Loctite® epoxy putty (Henkel Technologies, Rocky Hill, CT). Experiments were conducted at room temperature at 23 ± 1.5° C.

Figure 7.1 Pipe-drum apparatuses

Experimental Procedure

Gasketed/Ungasketed PVC Pipes Exposed to Gasoline

Eight pipe-drum apparatuses were constructed for this test, three gasketed PVC pipes with SBR gaskets (triplicate experiment), three gasketed PVC pipes with NBR gaskets (triplicate experiment), and two ungasketed PVC pipes as controls. To prepare a pipe-drum apparatus, silica sand was placed into the apparatus until the drum was nearly full. Premium gasoline was then added until a visible liquid level of gasoline appeared above the surface of the sand. The lid was secured to the pail with the lever-locks, minimizing the volatilization of gasoline. The PVC pipes were filled with deionized water with zero headspace. Pipe-water was removed each week for determination of BTEX compounds by gas chromatography, as previously described. The pipe was flushed with deionized water three times immediately after each sampling event before fresh deionized water was added and the needle valves were closed.

Gasketed PVC Pipes Exposed to Gasoline-Contaminated Water

To prepare gasoline-saturated water, a mixture of 1 L premium gasoline and 10 L deionized water was moderately stirred with a magnetic stirrer in a glass bottle for 48 hours. Several preliminary experiments indicated that a mixing time of 48 hours was sufficient to produce a saturated gasoline solution. The concentration of the dissolved BTEX was around 150 mg/L which was slightly higher than the average concentration of 135 mg/L reported by Cline et al. (1991) for water saturated with various grades of gasoline. Benzene and toluene were the major aromatic compounds in the aqueous phase, accounting for nearly 90% of total BTEX. Three saturated aqueous solutions of varying concentrations (75%, 50%, and 10% saturated aqueous solutions of gasoline) were prepared by diluting the stock solution with deionized water at appropriate ratios. In the experiments, a prepared gasoline solution was introduced into the drum through the copper tubes from bottom to top, using a Masterflex pump, until the drum was full.

The aqueous solutions in the drum were replaced with fresh aqueous gasoline solutions each month to maintain relatively constant exposure concentrations. BTEX concentrations in the drum were periodically measured.

RESULTS AND DISCUSSION

Gasketed/Ungasketed PVC Pipes Exposed to Gasoline

Pipes with SBR gaskets

The cumulative mass of BTEX compounds permeated over time are plotted in Figure 7.2. The triplicate experiments (SBR1, SBR2 and SBR3) demonstrated that the permeation of BTEX through gasketed PVC pipes with SBR gaskets was rapid when exposed to premium gasoline. Of the five compounds of interest, benzene was the first compound detected with a breakthrough time of about 21 days. When breakthrough occurred, the level of benzene in the pipe-water immediately exceeded the maximum contaminant level (MCL) of 5 μg/L. Toluene breakthrough lagged slightly behind that of benzene, while ethylbenzene and xylene were detected later at

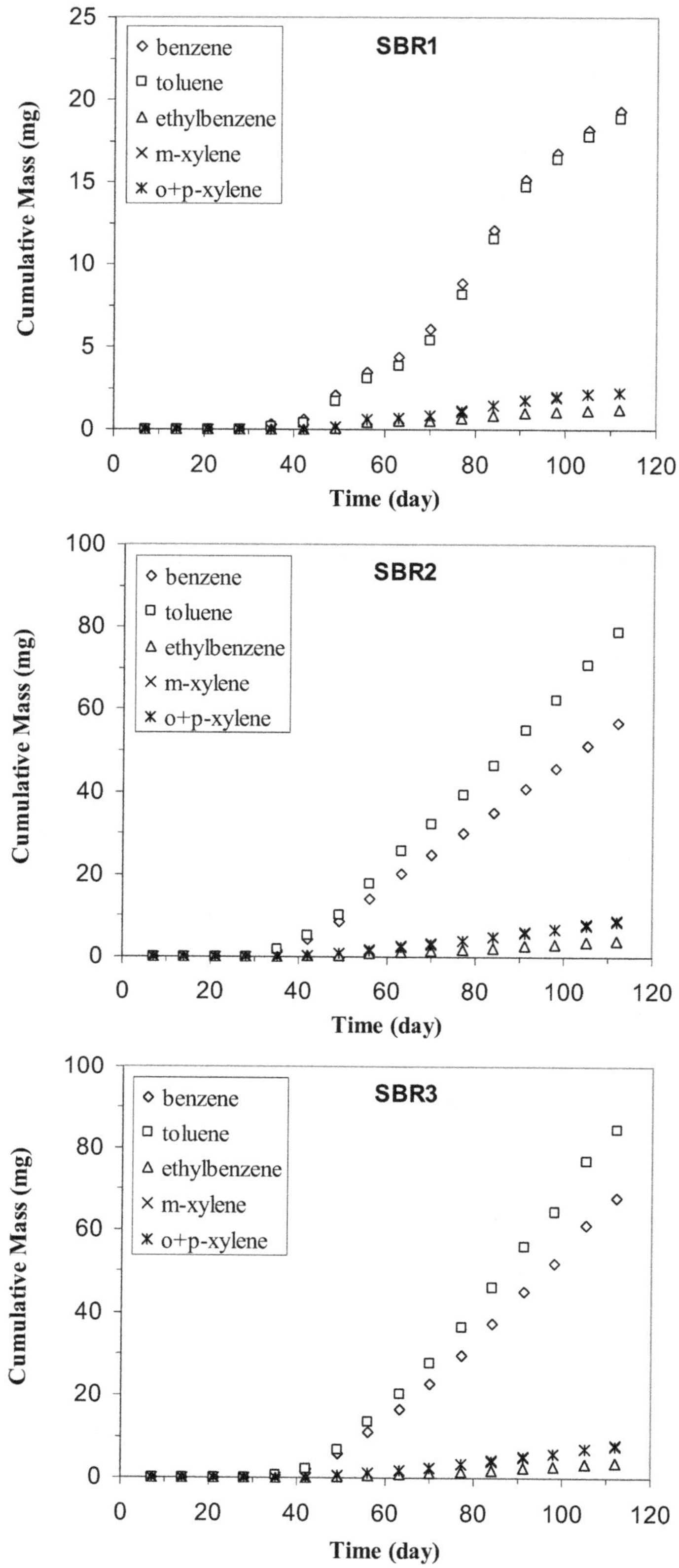

Figure 7.2 Cumulative mass of BTEX compounds permeated per joint in PVC pipes with SBR gaskets exposed to premium gasoline

about 50 days and at much lower concentrations. Benzene and toluene accounted for 85 - 90% of the entire total BTEX that permeated into the pipe-water. This is mainly due to the higher solubility of benzene in water and the higher mass fraction of toluene in gasoline. Similar permeation behavior for BTEX (i.e., the order of breakthrough time and the relative permeation rate) was found for the permeation of premium gasoline in PE pipes (Chapter 3) and ductile iron pipe joints (Chapter 6).

The results of the triplicate tests indicated that the cumulative mass of BTEX permeated varied for the three gasketed pipes under the same exposure conditions. The steady state permeation rates for benzene for SBR1, SBR2 and SBR3 were estimated to be 0.43, 0.76, and 1.0 mg/joint/day, respectively. The corresponding permeation rates for toluene were 0.45 mg/joint/day, 1.1 mg/joint/day and 1.3 mg/joint/day, respectively. Overall, the estimated permeation rates for SBR2 and SBR3 were similar to each other but the permeation rate for SBR1 was relatively lower. Factors resulting in the variability of permeated mass will be discussed later.

Pipes with NBR gaskets

The cumulative masses of BTEX permeated through NBR gaskets versus time are plotted in Figure 7.3. The triplicate experiments (NBR1, NBR2 and NBR3) demonstrated that permeation of BTEX through gasketed pipes with NBR gaskets was rapid under exposure to premium gasoline. The breakthrough time for benzene was 21 days. These results were different from the results for gasketed ductile iron pipe joints which showed NBR gaskets were more resistant to gasoline permeation than SBR gaskets. Using the sorption test with premium gasoline, the NBR rubber from Rieber gaskets used in PVC pipes gained about 75% of its original mass at equilibrium while the NBR rubber from Tyton® gaskets used in ductile iron pipe only gained 24% of its original mass. The results of the sorption test implied that the NBR materials for the PVC and ductile iron pipe were not identical in their compounding or their permeation properties.

The results of the triplicate tests for NBR gaskets also indicated that the mass permeated varied for the three NBR joints under the same exposure conditions. The steady state permeation rates for benzene for NBR1, NBR2 and NBR3 were estimated to be 0.22, 0.0008, and 0.34 mg/joint/day, respectively. The corresponding permeation rates for toluene were 0.56, 0.0008 and 0.56 mg/joint/day, respectively. Overall, the estimated permeation rates for NBR1 and NBR3 were similar to each other. The mass permeated for NBR2 was characterized by a constant but extremely low permeation rate.

Ungasketed Pipes

As expected, no BTEX compounds were detected in the pipe water of the ungasketed control experiments after 6 months of exposure.

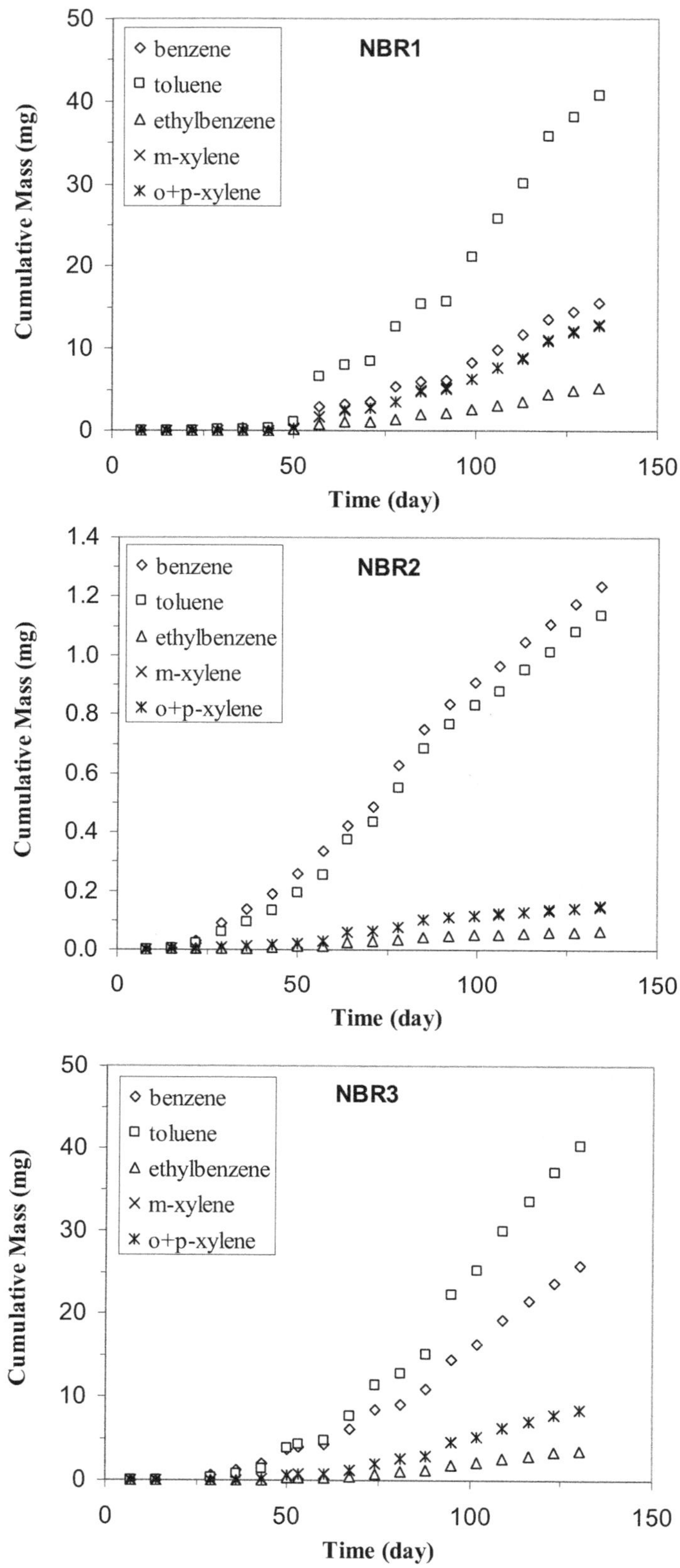

Figure 7.3 Cumulative mass of BTEX compounds permeated per joint in PVC pipes with NBR gaskets exposed to premium gasoline

Prediction of Benzene Concentration in Gasketed PVC Pipes Under Conditions of Stagnation

The benzene permeation rates presented above can be used to predict the concentration of benzene in the pipe water after a period of stagnation.

$$C_{pw} = \frac{M}{V} = \frac{n\ P_m\ t}{\frac{1}{4}\ \pi\ I.D.^2\ L} \tag{7.1}$$

where M = total mass of permeated contaminant (μg)
V = volume of water in the pipe (L)
n = number of pipe joints
P_m = steady state permeation rate of benzene (μg/day/joint)
t = period of stagnation (days)
L = length of contaminated pipelines (cm)
I.D. = inside diameter of pipe (cm)

To estimate the concentration of benzene for the worst-case scenario, the average permeation rate of benzene estimated for the experiments plus 3 standard deviations (i.e., 1.59 mg/joint/day for SBR and 0.70 mg/joint/day for NBR) were used for the calculations. Assuming a scenario of 100 ft of 2-inch PVC pipeline (including 5 gasketed joints) that was immersed in premium gasoline, the benzene concentrations in the pipe water after an 8 hour stagnation period would be 33 μg/L for SBR gaskets and 14.4 μg/L for NBR gaskets. In this scenario, use of both of SBR and NBR gaskets would have resulted in exceeding the 5 μg/L MCL in the pipe water.

Prediction of Benzene Concentration in Gasketed PVC Pipes Under Conditions of Full-Pipe Flow

Under normal conditions, water flows in the pipes and the contaminant concentration depends on the flow rate. In this case, assuming a continuous, full-pipe flow and complete mixing of permeated contaminants in the pipe water, the concentration of the contaminant in the pipe water (C_{pw}) can be estimated by:

$$C_{pw} = \frac{M}{V} = \frac{n\ P_m\ t}{q\ t} = \frac{n\ P}{q} \tag{7.2}$$

where q = water flow rate (cm^3/d)

Assuming the permeation rate for benzene is 1.59 mg/joint/day for SBR, 84.0 gal/joint/day or 318 L/joint/day of water would be required to dilute the concentration of benzene to the MCL of 5 μg/L. This corresponds to an average flow rate of 0.059 gpm in a 2-inch PVC pipe main. Since the analytical method detection limit (MDL) for benzene is 0.24 μg/L for this study, a flow of 1.24 gpm would dilute benzene permeating the SBR gasket in this worst case scenario to levels that are not analytically detectable.

Similar calculations were made for NBR gaskets. If the permeation rate of benzene is 0.70 mg/joint/day, a flow of 0.025 gpm is needed to dilute the benzene concentration to the MCL and a flow rate of 0.54 gpm to render the concentration in the pipe-water undetectable.

Typically, the average water flow velocity for pressure PVC pipes ranges from 0.6 m/s to 3.1 m/s (2 ft/s to 10 ft/s) (Uni-Bell 2001). For 2-inch PVC pipes, therefore, the corresponding water flow rates ranged from 25 gpm to 132 gpm. Due to the effect of dilution, the concentration of benzene in gasketed PVC pipes with a minimal average water flow would not exceed the MCL.

Discussion of Permeation Process in PVC Pipe Joints

Figure 7.4 shows the cross section of a pipe joint with a Rieber gasket. The permeation of contaminants through the PVC pipe joint will involve three processes: (i) contaminants in the environmental media diffuse/move through the gap between the bell and the spigot of the pressure pipe joints to reach the gasket; (ii) contaminants then diffuse through the gasket material; and (iii) the contaminants diffuse/move through the gap between the bell and the spigot of the pressure pipe joints reach the pipe water. The permeation of contaminants through pipe joints is not only dependent on the properties of gasket materials but also on the size of gap between the bell and the spigot. Generally, the gap size varies from joint to joint and is relatively small (less than 300 μm) and is mainly due to the axial joint deflection within allowable limits. Moreover, during the assembly of pipe joints, lubricant is applied to the spigot and the lubricated spigot is then pushed past the gasket into the bell. For the pipe joint shown in Figure 7.4, significant amounts of lubricant were found to occupy the empty space between the gasket and the spigot. The lubricant may affect the permeation of contaminants through the pipe joints by either covering the exposure area of gasket or providing another pathway for diffusion or a medium for sorption of the petroleum hydrocarbons. The effect of gap size and the amount of lubricant may account for the discrepancy detected in the testing of triplicate samples reported above.

Gasketed PVC Pipes Exposed to Gasoline-Contaminated Water

No significant amounts of BTEX compounds were detected in the pipe-water for pipes using either SBR or NBR Rieber gaskets after 9 months of exposure to any level of gasoline contamination in water, even 100% saturated. The concentrations of BTEX in many pipe water-samples were below the detection limits. In some pipe water samples, benzene and toluene were detectable (≥ 0.24 μg/L) but the concentrations were extremely low (much less than the MCL) and no permeation rate could be determined. The MCL for benzene will not be exceeded for any level of gasoline contamination in groundwater.

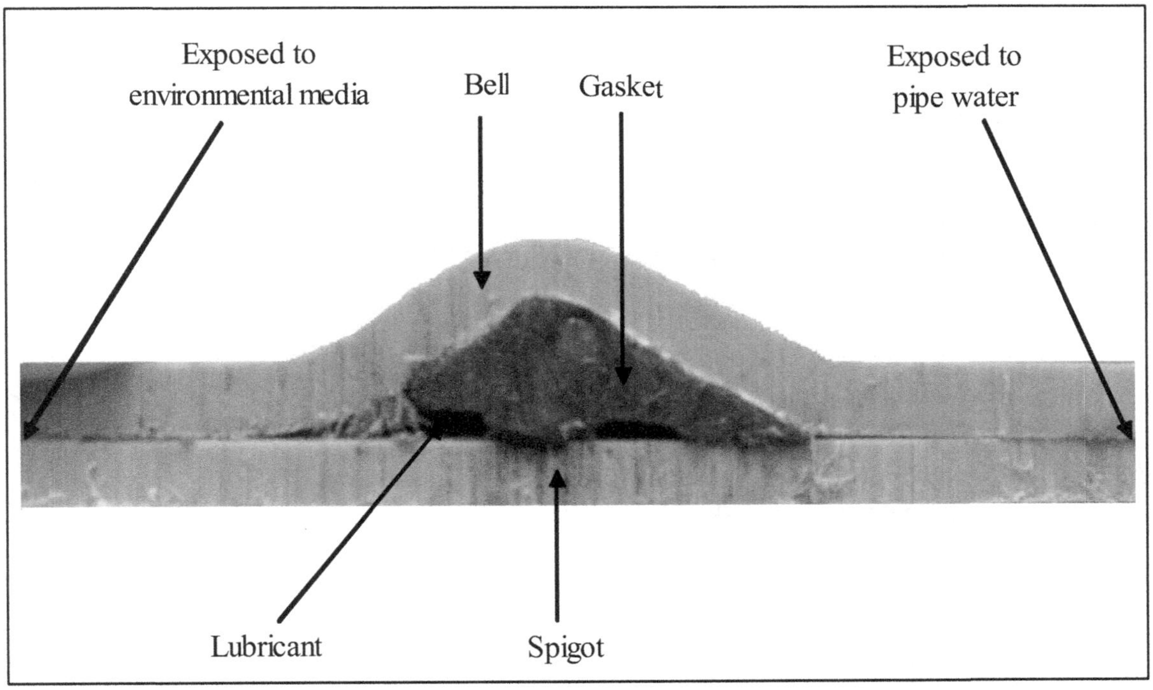

Figure 7.4 Cross section of PVC pipe joint with SBR Rieber gasket

Scale-up of Permeation in PVC Pipe with Rieber Gaskets

Equation 6.6 stated that

$$M_2 = M_1 \frac{L_1}{L_2} \frac{A_2}{A_1} \tag{6.6}$$

where M = mass permeated per joint per day (ug/joint/day)
L = length of permeation path (cm)
A = area of the gasket contacted by solvents (cm^2)

Since the ratios of the dimensions of the cross sections of Rieber gaskets are essentially identical for all pipe diameters (i.e., the cross sections are congruent), L and A are directly proportional to the diameters. Let

$$L = \theta H \tag{7.3}$$

where θ = a proportionality constant for all Rieber gaskets
H = length of any given dimension of the cross section of the gasket (cm)

Therefore,

$$\frac{L_1}{L_2} = \frac{\theta H_1}{\theta H_2} \quad (7.4)$$

where L_1 and H_1 refer to the gasket diameter for which M_1 was experimentally determined, and where L_2 and H_2 refer to the gasket diameter for which M_2 is to be estimated.

Let

$$A = \delta H d \pi \quad (7.5)$$

where δ = a proportionality constant for all Rieber gaskets
d = diameter of the gasket

Therefore,

$$\frac{A_2}{A_1} = \frac{\delta H_2 d_2 \pi}{\delta H_1 d_1 \pi} \quad (7.6)$$

Where d_1 = gasket diameter for M_1 was experimentally determined
d_2 = gasket diameter for which M_2 is to be estimated

Substituting Equations 7.4 and 7.6 into equation 6.6,

$$M_2 = M_1 \frac{\theta H_1}{\theta H_2} \frac{\delta H_2 d_2 \pi}{\delta H_1 d_1 \pi} = M_1 \frac{d_2}{d_1} \quad (7.7)$$

Hence, M can be estimated for gaskets of any diameter by using permeation data experimentally determined from gaskets of a single diameter and the ratio of the diameters, d_2/d_1, as a scaling factor.

Permeation rates of benzene through Rieber gaskets larger than 2-inch diameter were estimated using equation 7.7 and are presented in Table 7.1. The estimated permeation rates were based on the experimental permeation rates of benzene for a 2-inch PVC pipe from pipe-drum experiments, which were 1.59 and 0.70 mg/joint/day for SBR and NBR gaskets exposed to pure gasoline, respectively. It should be noted that the two permeation rates (1.59 and 0.70 mg/joint/day) were calculated as the mean value plus three standard deviation obtained from the triplicate experiments. As shown in Table 7.1, the permeation rate for a single joint increases with larger diameters, mainly due to their relatively large exposure area.

Figure 7.5 plots the flow rates needed to dilute concentration of benzene to the MCL in 2-inch to 24-inch PVC pipes with Rieber gaskets. For any size of PVC pipe (up to 24-inch) with SBR or NBR gaskets exposed to gasoline, a flow rate of 4 gpm is sufficient to dilute the contaminant level to less than the benzene MCL for 100 feet (5 joints) of pipe. As stated earlier, the flow velocities in water distribution pipes ranged from 2 to 10 feet per second (ft/s) (0.6 to 3 m/s). Using 2 ft/s, the flow rate for 2-inch pipe is estimated to be 25 gpm, which probably

Table 7.1
Estimated permeation of benzene in 4-inch to 24-inch PVC pipes with Rieber gaskets

Pipe Diameter (ins)	Permeation Rate (mg/joint/day)[1] SBR-Gasoline[2]	NBR-Gasoline[2]
2	1.59[3]	0.70[3]
4	3.18	1.40
6	4.77	2.10
8	6.36	2.80
10	7.95	3.50
12	9.54	4.20
14	11.1	4.90
16	12.7	5.60
18	14.3	6.30
20	15.9	7.00
24	19.1	8.40

[1]mean plus 3 standard deviations
[2]gasket in contact with pure premium gasoline
[3]from experimental data

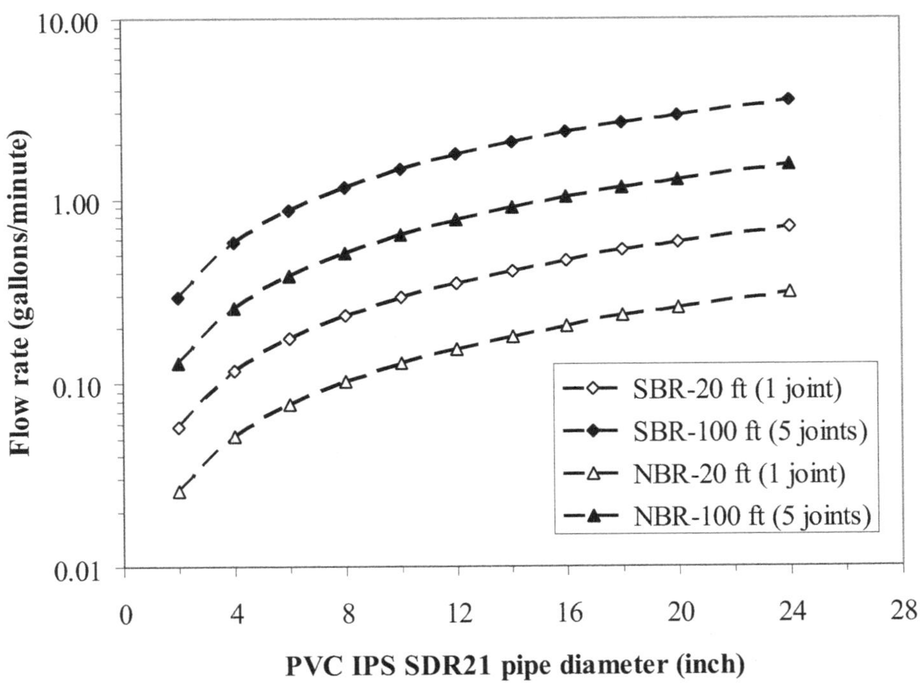

Figure 7.5 Flow rates needed to dilute concentration of benzene to MCL in 2-inch to 24-inch PVC pipes with Rieber gaskets.

represents the minimum flow rate encountered in most service conditions. This minimal flow will provide sufficient dilution so that benzene MCL will not be exceeded.

The concentrations of benzene that might result from periods of stagnation were estimated using benzene permeation rates in Table 7.1 and are presented in Figure 7.6. As shown in Figure 7.6, the benzene MCL would be exceeded for a 20-inch or smaller size pipe with SBR gaskets after 8 hours of water stagnation when pipes are exposed to free product gasoline. In the case of NBR gaskets, benzene MCL would be exceeded for a 10-inch or smaller size pipe under identical exposure conditions.

CONCLUSIONS

Pipe-drum tests were conducted to investigate the permeation of BTEX compounds through gasketed PVC pipes with Reiber gaskets under simulated field conditions. Simulated conditions included subsurface gasoline spills and gasoline-contaminated groundwater with various levels of contamination. Based on the analysis of experimental results, the following conclusions can be drawn:

(i) In comparison with ungasketed PVC pipes, the gasketed PVC pipes are much more vulnerable to permeation.

(ii) Under conditions of stagnation in contact with free product gasoline, the 5 µg/L MCL for benzene will likely be exceeded during an 8 hour stagnation period for a 20-inch or smaller size gasketed PVC pipe with SBR Rieber gaskets in contact with premium gasoline. The benzene MCL would be exceeded for a 10-inch or smaller size gasketed PVC pipe with NBR Rieber gaskets under the identical stagnation and exposure conditions.

(iii) Under conditions of continuous full pipe flow, the 5 µg/L MCL for benzene will not be exceeded even under the worst conditions of gasoline contamination (contact with free product), provided that there is at least a minimal flow of water in the main (0.037 gpm in 2-inch pipe).

(iv) No level of gasoline contamination in groundwater will cause the benzene MCL to be exceeded in PVC pipes equipped with SBR or NBR gaskets.

(iv) The permeation process of contaminants in PVC pipe joints may be affected by the use of lubricants during the pipe installation, and by the size of the gap between the bell and the spigot.

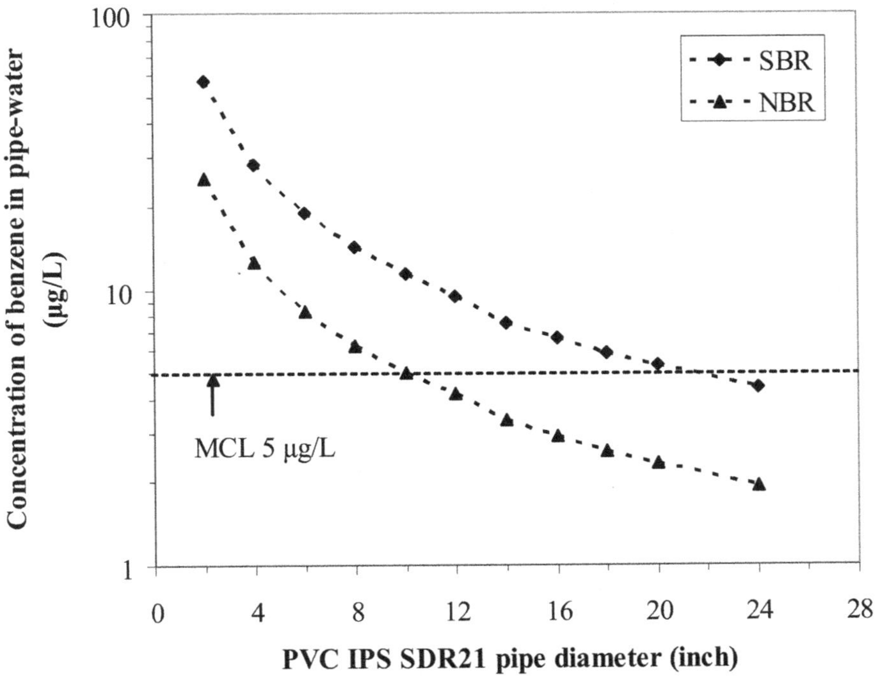

Figure 7.6 Benzene concentrations after 8 hours of stagnation in 100 feet of 2-inch to 24-inch PVC IPS SDR21 pipe with SBR and NBR Rieber gaskets in contact with gasoline.

CHAPTER 8
FINDINGS AND RECOMMENDATIONS

Table 8.1 provides a summary of the major findings in this report.

FIELD EXPERIENCES WITH HYDROCARBON PERMEATION (CHAPTER 2)

Water utilities in the U.S. and Canada were surveyed to obtain information regarding their practices for use of plastic pipes and gaskets and their experiences with permeation by hydrocarbons. Responses were received from 151 utilities with more than 83,000 miles of mains and nearly 5.5 million service connections. The responses were well distributed geographically and with regard to utility size. The case histories detailed in the Appendix were developed from these responses.

Eighteen percent of the total miles of mains reported were PVC, 16% were ductile iron, 66% reported others which included cast iron, steel, asbestos cement, and concrete pipes. Only 0.18% of the total mains reported were PE pipes. With respect to the number of service lines, PE and PVC pipe accounted for 6% and 5%, respectively.

The overall impact of permeation on the water industry is relatively small. Reports of permeation incidents involving potable water distribution are rare, with reports of one incident per 14,000 miles of mains and 0.9 incidents per million service connections in this survey. Nevertheless, every permeation incident that does occur is a potential threat to public health and customer satisfaction.

Twenty eight out of the 151 utilities (19%) reported having a protocol for selecting pipe and gasket materials for use in contaminated areas. Of the 28 utilities, 10 reported protocols specifying DI pipe with resistant gaskets (either NBR or FKM); 8 reported protocols specifying DI with no mention of gasket type; and 4 reported "no plastic", which could mean using DI in contaminated areas; 2 reported using steel, and 4 indicate other approaches. No utility reported specifying plastic pipes in contaminated areas.

Permeation incidents involving service lines are usually detected by customer complaints of bad taste and odor. The universally applied corrective action for service lines is replacement with copper pipes, which is appropriate since the MCL of 5 μg/L for benzene can be exceeded before gasoline odor is detected. It should be noted that for most service lines in the U.S., replacement and maintenance are the responsibility of the customer.

Since the main concern for the water utility is to fix the contamination problem as quickly as possible in order to minimize health risk and customer inconvenience, quantifying the soil and groundwater contamination that caused a permeation incident may not be a priority. Data obtained from the survey described in this report were not sufficient to draw conclusions regarding the thresholds of hydrocarbon contamination that would result in hydrocarbons permeation and require a utility to choose more resistant pipe or gasket materials.

Table 8.1 Summary of Tasks, Experiments, and Findings

Chapter	Description	Findings	Recommendations
Chapter 2	Survey and Case Histories	• Total miles of mains reported: 18% - PVC, 16% - ductile iron, 0.18% - PE, 66% - others (cast iron, steel, asbestos-cement, and concrete pipes). • No. of service connections reported: 6% - PE, 5% - PVC • One permeation incident per 14,000 miles of mains and 0.9 incidents per million service connections • 28 out of 151 utilities reported having a protocol for selecting pipe/gaskets for use in contaminated areas: 10 - DI pipe with resistant gaskets (either NBR or FKM); 8 - DI with no mention of gasket type; 4 - no plastic, 2 – steel, 4 – other approaches	
Chapter 3	Permeation of PE Pipes	• Gasoline permeates rapidly through PE pipes at all saturation concentrations and contaminated soils conditions • Permeation rate is a function of aqueous or vapor concentration of the hydrocarbons • Depending on hydrocarbon concentrations in soil water and soils and pipe size, water flow (no stagnation) may dilute benzene concentration to below MCL	• No level contamination at which PE pipe is resistant to permeation of gasoline, chlorinated solvents or aqueous solutions of gasoline • Corrective action should be taken
Chapter 4	Permeation of Ungasketed PVC pipes	• Premium gasoline (and, therefore, groundwater or soils saturated with gasoline) - no permeation of PVC pipes • Pure toluene, benzene and TCE – permeation of PVC pipes • saturated vapors of toluene, benzene and TCE – permeation of PVC pipes • Saturated aqueous solutions of benzene, toluene, or TCE – permeation of PVC pipe if activity > 0.60	• PVC pipes impervious to gasoline and gasoline-saturated solutions • PVC pipes permeated by saturated aqueous solutions of benzene, toluene, and TCE • Decontamination of PVC not feasible – must replace permeated pipe
Chapter 5	Predictive Tests (Sorption, Moving Front and Near Infrared (NIR) Reflectance)	• Of the three predictive tests (sorption, moving front and near infrared spectroscopy), the moving front test is the most suitable for PVC pipes and provides information on permeation rates • For single compound concentrations $\leq$40% (v/v) of benzene, toluene, or TCE in reference fuel, sorption (permeation) was insignificant and moving fronts were not formed. • No moving fronts were formed in $\leq$ 60% (v/v) saturated aqueous solutions of benzene, toluene, or TCE • Premium gasoline fortified with toluene - no moving front observed for $\leq$ 32.2% (v/v) total BTEX, moving fronts formed for $\geq$ 43.1% (v/v) total BTEX • No moving fronts for $\leq$40% (v/v) of multi-component BTEX mixtures in reference fuel. Fronts formed for $\geq$ 60% (v/v)	• PVC pipes can tolerate an activity of 0.40 (40% aqueous solutions of benzene, toluene and TCE) without permeation (without including safety factor and impact of other environmental conditions) • PVC pipes impervious to dilute aqueous solutions of benzene, toluene, and TCE, commonly encountered in the field • PVC pipes compatible with sum of individual BTEX activities $\leq$0.40 (activities are additive)

(continued)

Table 8.1 Summary of Tasks, Experiments, and Findings (Cont'd.)

Chapter	Description	Findings	Recommendations
Chapter 6	Permeation of Gasketed (Tyton®) DI Pipe	• Gaskets exposed to premium gasoline: EPDM permeated in 35 days, CR in 50 days, SBR in 50 days (4", benzene – avg. 2.72 mg/j/d), NBR in 124 days (4", benzene – 0.159 mg/j/d) (FKM (very low BTEX detected – probably leakage) • SBR gaskets permeated when exposed to gasoline-saturated water (210 days) and 50% saturated (240 days) • No permeation observed for SBR gasket in 20% and 5% saturated aqueous solutions of gasoline • No permeation of NBR gaskets observed for any level of gasoline contamination in water • No correlation between permeation and hydrostatic pressure in pipe	• SBR and NBR gaskets may be used (without exceeding benzene MCL) under all conditions of gasoline contamination with minimal average flow in pipe (no stagnation) • Under conditions of stagnation in contact with free product gasoline, NBR gaskets may be used (without exceeding benzene MCL). • SBR and NBR gaskets compatible with any level of gasoline in groundwater • NBR gaskets most cost effective choice when a gasket material resistant to gasoline is desired.
Chapter 7	Permeation of Gasketed (Rieber) PVC pipes	• Permeation of premium gasoline observed within 21 days for SBR (2", benzene avg. 0.73 mg/j/d) and NBR gaskets (2", benzene avg. 0.19 mg/j/d). • No permeation after 9 months for SBR and NBR gaskets in gasoline-saturated water • NBR rubber compound used for Rieber gaskets is about 3x more permeable than NBR rubber compound used in Tyton gaskets (75 W% vs 24% W% sorption).	• SBR or NBR gaskets may be used (without exceeding benzene MCL) in direct contact with gasoline with a minimal average flow in pipe • After 8 hours of stagnation in direct contact with gasoline, benzene MCL exceeded in pipes ≤ 24" using SBR and pipes ≤ 10" using NBR gaskets. • SBR or NBR gaskets compatible with any level of gasoline in groundwater (without exceeding MCL)

HYDROCARBON PERMEATION OF PE PIPE (CHAPTER 3)

Laboratory Experiments

Laboratory experiments indicated that HDPE pipe is rapidly permeated by gasoline and solvents, whether in groundwater, in soil pore vapor, or as the free product. The rates of diffusion of the BTEX compounds in gasoline through HDPE pipe are logarithmically dependent on the external bulk concentration and are an order of magnitude faster for free product gasoline than for saturated groundwater or soil pore vapor (diffusion coefficients are 10^{-8} cm^2/s and 10^{-9} cm^2/s, respectively).

Benzene was the first of the BTEX compounds to permeate HDPE pipe from water contaminated with gasoline, with the first detectable concentrations exceeding the 5 μg/L MCL for all experiments. The odor threshold concentration for benzene varies from 23 μg/L (US EPA, 2006) to about 190 μg/L (Young et al. 1996). Since most permeation incidents involving service lines were identified as a result of customer complaints of bad taste and odor, it is likely that the concentration of benzene in drinking water exceeded the EPA MCL in all incidents.

Equations (Equations 3.5 and 3.6) were derived to predict the concentration of contaminants in PE pipes for various external contaminant concentrations during continuous flow and after periods of stagnation. For a given contaminant in soil water, smaller pipes are more vulnerable than larger pipes and the risk of exceeding EPA MCLs is much higher during periods of stagnation than during continuous flow.

Laboratory experiments showed that for a given mass of contaminants in soil, organic matter in the soil strongly sorbed the contaminants, decreasing the contaminant concentrations in soil water or soil vapor, thereby decreasing the permeation of contaminants into HDPE pipe. While the organic content of soils can effectively lower the concentration of hydrocarbon contaminants in the soil pore water, once the organic adsorption capacity of the soil is exceeded, the organic content of the soil makes no difference. This indicates that the steady state permeation rates in HDPE pipes are dependent on the external bulk concentrations regardless of the soil types. While organic content might reduce permeation incidents in the case of transient hydrocarbon contamination, it would be irrelevant for continuous contamination.

Recommendations for PE Pipe

For practical purposes, there is no level contamination at which PE pipe is resistant to gasoline or chlorinated solvents. Immediate corrective action should be taken whenever there is a known spill of gasoline or organic solvents in the vicinity of a PE service line. It is important to note that permeation occurs regardless of external concentration. Regardless of whether there is a taste or odor complaint, it is most likely that the MCL for benzene will be quickly exceeded in the potable water under conditions of stagnation as the taste and odor threshold level for gasoline in potable water appears to be well above the MCL for benzene. However, under full flow conditions, the benzene concentration in pipe water may not exceed the MCL depending on the external hydrocarbon concentrations, the pipe size and exposed surface area of the pipe.

HYDROCARBON PERMEATION OF PVC PIPE (CHAPTER 4 and CHAPTER 5)

Laboratory Experiments – Pipe-Bottle Apparatus (Chapter 4)

Pipes used in the pipe-bottle experiments were new pipes purchased from various building supply stores. The pipe-bottle experiments were conducted over a two-year period at room temperature ($23 \pm 1.5\ ^{\circ}$C).

Pipe-bottle experiments showed that PVC pipe was rapidly permeated by pure benzene (20 days), toluene (16 days), and TCE (6.5 days). PVC pipe was also rapidly permeated by saturated benzene vapor (31 days), toluene (28 days), and TCE vapor (13 days). The activities of pure solvents are equal to one. Similar behavior can be expected from closely related swelling compounds such as ethylbenzene, xylenes, tetrachloroethylene (PCE), 1,1-dichloroethylene (1,1-DCE), and 1,2-dichloroethylene (1,2-DCE). Pipe-bottle experiments also indicated that saturated aqueous solutions of TCE and benzene permeated the PVC pipes at 168 and 250 days, respectively, and that permeation by saturated aqueous solutions of toluene was likely to occur in about 12 months.

Pipe-bottle experiments indicated that PVC pipes were impervious to gasoline and gasoline-saturated water over two years of exposure and, therefore, can be used in soils contaminated with gasoline, regardless of the level of contamination. An explanation for this resistance to permeation is that the premium gasoline did not have sufficient swelling compounds such as benzene, toluene and xylene to soften the PVC material. The activity of total BTEX in gasoline for the pipe-bottle experiments was approximately 0.25.

Experiments indicated that permeation of PVC pipe can not be readily reversed by soil remediation and flushing of the pipe. PVC pipes permeated with benzene continued to leach benzene into the pipe-water for six months after external contamination was removed (at which time the experiment was terminated). While it seems likely that permeated PVC pipe would bust due to loss of strength, any PVC pipe that does remain intact after permeation should be replaced as remediation is not feasible.

Laboratory Experiments – Predictive Tests for PVC Pipe (Chapter 5)

Gravimetric Sorption Test

The gravimetric sorption test is an indirect measure of permeation in a PVC pipe by measuring the weight gain of the test specimen when exposed to gasoline or solvents. Using this test, the plot of the sorption rates (W%/day$^{1/2}$) and the % by volumes of toluene in polyethylene glycol (PEG, an inert chemical) showed that for 30% by volume of toluene and below, sorption of toluene was insignificant – implying that no permeation of toluene occurs below this concentration. With aqueous saturated solutions of benzene, toluene, or TCE, the weight gain for aqueous solutions of $\leq$ 60% saturation was less than 2% after 8 months of exposure. Sorption from the 20% aqueous solutions of toluene and benzene was found to be statistically similar to that of the control experiment - implying that sorption was negligible for contaminant levels below 20% of aqueous saturation for both toluene and benzene. Solutions at 20% aqueous saturation (toluene and benzene concentrations at 100 mg/L and 340 mg/L, respectively) are considered to be high levels of environmental pollution which are seldom encountered in the field except in close proximity to a NAPL.

The rate and extent of sorption of toluene from NIST reference fuel solutions by PVC pipe was considerably lower than from PEG solutions. NIST reference fuel is seen as a more relevant surrogate for gasoline than PEG because it is chemically similar to the non-aromatic compounds in gasoline. There was significant sorption of toluene from reference fuel at toluene concentrations of $\geq$ 60% (v/v) and no significant sorption at concentrations of $\leq$ 40% (v/v).

Sorption of premium gasoline was found to be negligible with less than 1% weight gain – implying that no permeation occurred. The gravimetric sorption test results showed that the rates of sorption of solvents parallel the progress of the moving front test described below.

Moving Front Test

The moving front test is a precise and rapid test that directly measures the progression of permeation caused by swelling in a PVC pipe. Reflected light microscopy is used to visualize and measure the rate of advance of the moving front over a 48 hour test period. The moving front test can be used to quantify the resistance of PVC pipe to permeation.

Moving front rates for 58 specimens of PVC pipes of differing lot codes from five different manufacturers and diameters (0.5 ins. to 2 ins.) in a 48-hour toluene test were found to range from 0.181 to 0.236 mm/hour$^{1/2}$. This result showed that the moving front test can be used to compare the relative susceptibility of PVC pipes to permeation.

In experiments with PVC pipes exposed to various volume percents of toluene in PEG, no moving front was found when the volume percent of toluene was less than 25%. In experiments with PVC pipes exposed to various concentrations of toluene in the reference fuel, no moving front was found when the volume percent of toluene was 40% (v/v) or below. For aqueous solutions of benzene, toluene or TCE, moving fronts were observed at 100% and 80% of saturation, but no moving fronts at 60%, 40%, or 20% of saturation after 12 months of exposure.

Moving fronts did not form during two years of exposure of PVC pipes to pure gasoline or to aqueous solutions of gasoline after one year of exposure. Premium gasoline fortified with toluene did not cause a moving front to form in PVC pipe when the volume percent of total BTEX and toluene were at or below 32.2% and 21.2%, respectively. Moving fronts were formed when total BTEX was between 43% and 45% by volume and toluene was at or above 30.8% by volume, a concentration of toluene that is more than double that typically found in premium gasoline and much higher than in regular or E10 gasoline. These results, plus those involving toluene in PEG and in reference fuel, are consistent with the sorption experiments indicating that PVC pipe is impervious to gasoline - even to premium gasoline, which has the highest concentrations of toluene and total BTEX of any commonly used motor fuel.

Moving fronts formed in PVC pipe when exposed to solutions of benzene, toluene, or TCE in NIST reference fuel of concentrations of 80% (v/v) and 60% (v/v), but did not form for concentrations of 40% (v/v) or 20% (v/v). Experiments with equal-volume-mixtures of benzene, toluene, ethylbenzene, and xylene in reference fuel showed that the contribution of BTEX compounds to initiating a moving front in PVC is additive. A 40% (v/v) solution of the mixture (i.e., 10% (v/v) of each of the four compounds) or lower did not initiate a moving front, but a 60% (v/v) solution (i.e., 15% (v/v) of each compound) or greater did initiate a moving front.

The moving front tests showed that premium gasoline did not have sufficient toluene or other BTEX compounds to initiate a moving front. The BTEX concentrations (v/v) for the premium gasoline used in this work were: toluene - 11.2 ± 1.9%, benzene - 2.8 ± 0.4%; ethyl benzene - 1.6 ± 0.2%; total (o-,m-,p-) xylenes - 8.4 ± 1.4%; total BTEX - 24.0 ± 3.9%. Experiments with toluene in PEG or toluene-fortified gasoline showed that a concentration of at least 30% (v/v) toluene or at least 40% (v/v) total BTEX is required to initiate a moving front. This 40% (v/v) threshold of concentration for initiating a moving front is consistent with sorption and moving front experiments using NIST reference fuel. This is well above the concentration of toluene in premium gasoline.

Near Infrared Reflectance Spectroscopy

NIR spectroscopy was found to be a useful screening tool for predicting pipe permeation susceptibility with the least weight regression (LWR) model by using the NIR spectra of unexposed pipes and the advance of the moving front as the laboratory reference data. However, the models developed are limited to the pipes that belong to pipe types used in the calibration population.

NIR spectroscopy is also a suitable tool for tracking the permeation process of pure solvents such as benzene and toluene in PVC pipes by using the NIR spectra of exposed pipes and laboratory reference datasets such as percent weight gain, weight gain per unit length and advance of the moving front.

Although permeation of gasoline into PVC pipes was negligible, the NIR spectra were found to provide a satisfactory degree of correlation with the length of time a specimen has been exposed to gasoline. Similarly, the number of days of exposure to aqueous saturated solutions of toluene and benzene were found to provide a satisfactory degree of correlation with the NIR spectra. However, the lower the aqueous saturation concentration, the higher was the prediction error.

The experimental results imply that NIR spectroscopy can be used as a predictive test tool once a database of reference laboratory data is established for a given model of NIR spectrometer with updating of the calibration set when new types of pipes are used or introduced. Predictions using the model, however, are limited to the types of PVC pipes within the database. NIR may not be able to make predictions with certainty for the types of PVC pipes not in the database.

Relevance of Predictive Tests

The moving front test, gravimetric sorption test, and NIR spectroscopy are useful for quantifying the resistance of PVC pipe to permeation. The moving front test is applicable only to PVC pipe because it depends on the formation of a visible moving front, which is the boundary between the rubberized and non-rubberized zones of the polymer. The moving front test can be used by utilities to assess the relative resistance of PVC pipes to permeation by hydrocarbons and as a specification for pipe performance. The moving front test could be incorporated into ASTM and other standards that specify performance criteria for PVC pipes. It is a direct and visible measurement of permeation of PVC pipes and can be used to test new as well as used pipes. The moving front test is a destructive test and requires the availability of a small specimen that can be sacrificed for testing. It would be a preferred means for studying the impact on PVC pipe by compounds not addressed in this study, such as light aromatics and chlorinated solvents other than benzene, toluene, and TCE, as well as ketones used in lacquer thinners and other commercial solvents.

The gravimetric sorption test is applicable to PVC as well as PE and rubber gaskets, which do not form visible moving fronts as permeation progresses, and would be applicable to compounds not included in this study.

In principle, NIR spectroscopy can be used to measure any physicochemical variable in an organic material that can be measured by an independent physical or chemical test. NIR is applicable to studying the interaction of PVC with compounds not included in this study and possibly could be used for studying permeability of and permeation in PE pipes and rubber gaskets. NIR spectroscopy holds promise for non-destructive testing to assess the condition of installed PVC pipe. Studies should be undertaken to develop and calibrate field-portable NIR instrumentation to measure brittleness, contaminant exposure history, or other parameters of interest to the water industry that are dependent on the physicochemical properties of the pipe material.

Recommendations for PVC Pipe

PVC pipe is impervious to gasoline. Where gasoline is the only source of hydrocarbon contamination, and a choice is to be made between PVC or DI pipe, engineering considerations other than permeation should govern the selection of pipe materials. However, if conditions of stagnation and contact with free product are anticipated, selection of DI pipe with NBR gaskets is preferable, as discussed below.

PVC pipe is impervious to benzene, toluene, and TCE at concentrations in water that are commonly encountered in the field. Based on the laboratory experiments, moving fronts were not observed for aqueous solutions with less than ≤ 60% (v/v) saturation of benzene, toluene, and TCE but weight gains of 2% in the sorption test were observed for 60% aqueous solutions of toluene and TCE. These levels of contamination are rarely encountered in the field and then only in areas of the most severe contamination. Similar light aromatic compounds (i.e., ethylbenzene and xylenes) and chlorinated solvents (i.e., PCE, DCE, PCA, and TCA) will likely behave in a similar manner, but further work would be needed to prove this conclusively.

AWWA standards relating to plastic pipes or gasketed pipes include a statement regarding permeation by hydrocarbons directing utilities to consult with manufacturers before selecting pipe or gasket materials to be used in contaminated or potentially contaminated areas. PVC pipe manufacturers commonly refer customers to the Uni-Bell PVC Pipe Association for guidance regarding use of PVC pipes in contaminated areas. The current criteria for compatibility with PVC pipe recommended by the Uni-Bell PVC Pipe Association (based on publications by Vonk (1985) and Behrens (1985)) are 0.25 activity for individual BTEX compounds and 0.10 activity for individual chlorinated solvents. These criteria were based on sorption experiments with permeation involving exposure of pure PVC polymer powders to solvent vapors and exposure of PVC sheets pressed from pure powders to the liquid solvents. The laboratory results here indicate that PVC pipes can tolerate a higher activity than that currently specified by Uni-Bell.

The sorption and moving front experiments using NIST reference fuel indicated that the contributions of the activities of individual BTEX compounds to the initiation of a moving front in PVC pipe are additive and do not depend of the activity of a single compound exceeding the threshold level. This conclusion was supported by the moving front experiments with toluene-fortified gasoline. The sorption and moving front experiments with aqueous solutions of benzene, toluene, and TCE indicate that an activity of greater than 0.60 (60% of maximum solubility) in a single-compound solution is required for initiation of a moving front in PVC pipe by any of those compounds. The authors recommend that utilities consider PVC pipe compatible with groundwater for which the sum of the activities of the individual BTEX compounds is ≤ 0.40, whether the contamination is from gasoline or other sources, provided that no other swelling solvents (such as ketones or chlorinated solvents) are present in significant amounts. The ≤ 0.40 activity limit also applies to TCE in groundwater, provided that it is the only swelling solvent present in a significant amount. An additive compatibility limit of ≤ 0.40 may also apply to the other chlorinated solvents not tested in this study, as well as ketones from laquer thinners, etc., but further work would be required to prove this conclusively. While the data indicate that an additive compatibility limit of 0.60 activity might be appropriate, the authors recommend 0.40 to allow for a safety factor.

HYDROCARBON PERMEATION OF GASKETS (CHAPTER 6 and CHAPTER 7)

Laboratory Experiments with Ductile Iron Pipes and Tyton® Gaskets (Chapter 6)

Gravimetric sorption tests for five different Tyton® gasket materials (EPDM, SBR, CR, NBR and FKM) for DI pipes in premium gasoline indicate that EPDM showed high sorption (127% weight gain) in comparison to the other gaskets, while FKM, known for high resistance to gasoline, exhibited very low sorption (about 0.65%).

Tyton® gaskets consist of a hard "heel" and a soft "bulb." Gravimetric sorption tests for the heel and bulb of the five gasket materials indicate that the bulb sorbed more than the heel for all gasket materials except for CR. For SBR Tyton® gaskets, the bulb sorbed about 66% more than the heel. The heel and bulb rubber consist of the same basic types of rubber but the heel is compounded with extra carbon black to make it harder (Bird 2006).

While the bulb rubber is known to provide the hydrostatic seal for the water in the pipe, observations indicated that an external contaminant must first pass through the less permeable heel rubber. The permeability of the heel rubber would, therefore, determine the overall permeability of an SBR gasket. This would not be the case for NBR gaskets since equilibrium sorption experiments showed the heel and bulb rubbers to be nearly identical.

Experiments with 4-inch DI joints using the five different Tyton® gasket materials exposed to free product premium gasoline at 0 psig showed that EPDM and CR performed similarly to SBR with equilibrium permeation rate of 5.20 total BTEX/joint/day for SBR. Breakthrough of gasoline was observed for the three gaskets after 35 days of experimentation. NBR (124 days) and FKM were the most resistant to permeation with equilibrium permeation rates of 0.36 and 0.49 (possible leakage) mg total BTEX/joint/day, respectively. These pipe-drum experiments indicated that NBR and FKM were somewhat more resistant to permeation than SBR.

After 7 months of exposure of SBR, NBR, and FKM, 4-inch Tyton® gaskets to saturated solutions of gasoline in water (about 168 mg/L total BTEX), breakthrough was observed through the SBR gasket after 210 days at a rate of 0.073 mg/joint/day of benzene (0.203 mg/joint/day of BTEX). No permeation was observed through the NBR or FKM gaskets exposed to saturated solutions of gasoline for 550 days (18 months). Permeation was observed for the SBR gasket exposed to 50% aqueous solutions of gasoline after 240 days. In addition, no permeation was observed through NBR gaskets exposed to 50%, 20%, and 5% saturated aqueous solutions of gasoline and for SBR gaskets exposed to 20%, and 5% aqueous solutions of gasoline.

Experiments with 4-inch DI pipe joints using SBR Tyton® gaskets exposed to free product premium gasoline, and under hydrostatic pressures of 0, 20, 40, and 60 psig indicated that there was no correlation between permeation and hydrostatic pressure in the pipe.

Under conditions of stagnation, such as in a service line, the 5 μg/L MCL for benzene will likely be exceeded during an 8-hour stagnation period for SBR gaskets in contact with free product gasoline or if they are impacted by benzene vapor from a free product layer in close proximity to the pipe. Under these circumstances, NBR gaskets would be sufficiently resistant to prevent permeation by benzene or other BTEX compounds in gasoline to a level that would exceed EPA MCLs. SBR Tyton® gaskets are sufficiently resistant to gasoline in groundwater, even groundwater that is saturated with gasoline, to prevent EPA MCLs from being exceeded.

Provided there is at least a minimal flow of water in the main, benzene and other BTEX compounds in gasoline that would not exceed EPA MCLs, even under the worst conditions of gasoline contamination, due to dilution.

Laboratory Experiments with PVC Pipes and Rieber Gaskets (Chapter 7)

Permeation of gasoline through 2-inch PVC pipes with SBR and NBR Rieber gaskets was rapid with breakthrough times for benzene of about 21 days. When breakthrough occurred, the level of benzene in the pipe-water immediately exceeded the maximum contaminant level (MCL) of 5 μg/L. The results were different from the results for gasketed ductile iron pipe joints which showed NBR gaskets to be more resistant to gasoline permeation than SBR gaskets.

For 8 hours of water stagnation in 2-inch pipe, benzene concentrations in the pipe-water for SBR and NBR gaskets were estimated to be 33.0 μg/L and 14.4 μg/L, respectively. In this scenario, use of either SBR or NBR gaskets in free product gasoline would have exceeded the 5 μg/L MCL for benzene.

Using the experimental permeation rate of 1.59 mg/joint/day for benzene through the SBR gaskets, an average flow rate of 0.059 gallons per minute (gpm) in a 2-inch PVC pipe main would be needed to dilute the concentration of benzene to the MCL of 5 μg/L. For NBR gaskets, at a permeation rate for benzene of 0.70 mg/joint/day, a flow of 0.025 gpm is needed to dilute the benzene concentration to the MCL. Typical flow rates in 2-inch pipe range from 25 gpm to 132 gpm at water flow velocities of 0.6 m/s to 3.1 m/s (0.2 ft/s to 10 ft/s). Therefore, in typical gasketed PVC pipes with full water flow, the concentration of benzene would not exceed its MCL due to the effect of dilution.

The benzene MCL would be exceeded for 20 in. or smaller pipes with SBR gaskets after 8 hours of water stagnation when pipes are exposed to free product gasoline. In the case of NBR gaskets, the benzene MCL would be exceeded for 10 in. or smaller pipes under the identical exposure conditions.

For any size of PVC pipe (up to 24 in.) with either SBR or NBR gaskets exposed to gasoline, a flow rate of 4 gpm is sufficient to dilute the contaminant level to benzene MCL for 100 feet (5 joints).

After six months of testing, there was no permeation of BTEX in either SBR or NBR Rieber gaskets exposed to any aqueous solutions of gasoline, including gasoline-saturated water.

Recommendations

Tyton® Gaskets

SBR Tyton® gaskets may be safely used (without exceeding the MCL for benzene) in DI water mains under all conditions of gasoline contamination as long as there is at least minimal average flow in the main (at least 0.24 gpm/joint in 4 in. pipe) and conditions of stagnation do not occur. Under conditions of stagnation, as in a service line in direct contact with free product gasoline or soil vapor from a nearby layer of free product, NBR gaskets may be safely used (without exceeding the MCL for benzene).

NBR Tyton® gaskets are the most cost effective choice when a gasket material resistant to gasoline is desired. NBR Tyton® gaskets are about 15 times more resistant to permeation by BTEX in gasoline than SBR Tyton® gaskets and cost about four times more than SBR gaskets.

FKM gaskets cost about 50 times more than SBR and it is not clear whether their performance in comparison to NBR is worth the cost.

Rieber Gaskets

Both SBR and NBR Rieber gaskets may be safely used (without exceeding the MCL for benzene) in PVC water mains in direct contact with free product gasoline provided there is a minimal average flow (at least 0.11 and 0.05 gpm/joint for 4 in. SBR and NBR, respectively). For conditions with 8 hour stagnation, the MCL will be exceeded $\leq$ 20-inch pipes with SBR gaskets and $\leq$ 10-inch pipes with NBR gaskets when gasketed pipes are exposed to free product gasoline. Both SBR and NBR Rieber gaskets may be safely used (without exceeding the MCL for benzene) for PVC pipes exposed to any level of gasoline contamination in groundwater.

CHAPTER 9
GUIDANCE TO WATER UTILITIES REGARDING PERMEATION BY HYDROCARBONS

INTRODUCTION

This chapter offers guidance to water utilities regarding the selection of pipe and gasket materials for use in areas contaminated by or at risk of contamination by certain petroleum-based hydrocarbons, and actions to be taken upon discovery that those hydrocarbons have come into contact with installed water distribution pipes.

This guidance applies specifically to the hydrocarbon contaminants studied in this work; i.e., gasoline, the most commonly encountered environmental contaminant that presents a permeation risk, as well as the solvents benzene, toluene, and trichloroethylene (TCE). Current AWWA standards contain a statement on permeation that directs users of the standards to "consult with the manufacturer regarding permeation." Nothing in this chapter should be construed to recommend not following the guidance in the AWWA permeation statement for contaminant-material combinations not studied in this report.

The guidance offered here applies specifically to risk of permeation of PE and PVC pipes and pipe gaskets by hydrocarbons. There may be other engineering considerations and regulatory requirements that govern the selection of materials for use in water distribution systems and remedial actions in the event of contamination. The guidance offered here is stated with a minimum of detail and supporting documentation. Readers interested in the justifications for this guidance are referred to the pertinent sections of this report.

This guidance is based on the assumption that water is safe to drink if U.S. EPA maximum contaminant levels (MCLs) are not exceeded for any contaminant. Not exceeding the 5 μg/L MCL for benzene is the basis for the guidance offered here regarding gasoline contamination. Sufficient information is given in this report for utilities to perform the equivalent calculations for other contaminants or lower contaminant levels. It should be noted that the EPA's maximum contaminant level goal (MCLG) is 0 μg/L.

PVC pipe refers to unmodified PVC (PVC-U) unless otherwise stated. Gasoline refers to premium gasoline, which has more toluene and total BTEX than other commonly available motor fuels. Premium gasoline was used in this work because it represents a worst-case scenario for permeation by motor fuels.

MAINS AND GASOLINE

PVC and DI Water Mains

Either PVC pipe with SBR Rieber gaskets or DI pipe with SBR Tyton gaskets may be used (without exceeding the MCL for benzene) for water mains in all conditions of gasoline contamination provided there is at least a minimal flow in the main. PVC pipe itself is impervious to gasoline. All gasoline permeation takes place through the gaskets for both PVC and DI pipe. For the worst case of gasoline contamination, immersion in free product gasoline, the flows necessary to avoid exceeding the 5 μg/L MCL for benzene are ≥ 0.24 gpm in a 4-inch DI pipe and 0.11 gpm in a 4 in. PVC pipe for each joint so impacted. SBR gaskets can be used safely in water mains in contact with groundwater contaminated with any level of gasoline, including saturation (about 70 mg/L benzene), provided there is a flow equivalent to at least 0.01 gpm in a 4 in. DI pipe for each joint so impacted.

PE Water Mains

Permeation of hydrocarbons through PE water mains depends on the concentration of contaminant in contact with the pipe, the length of pipe impacted, the formulation of the polymer, and the dimension ratio of the pipe. The tables and equations presented in Chapter 3 can be used to estimate contaminant concentration under periods of stagnation or flow to formulate site-specific decisions regarding installation of PE mains in contaminated or potentially contaminated areas as well as decisions regarding remediation of installed mains impacted by hydrocarbon contamination.

SERVICE CONNECTIONS AND GASOLINE

PVC Service Connections

As described above, PVC pipe is impervious to gasoline, either the free product, the vapor, or in groundwater. Small PVC service lines are commonly joined by cementing. The permeability of cemented PVC joints was not investigated in this study.

PVC service lines (2-inch and larger) joined with SBR and NBR Rieber gaskets can be used in groundwater contaminated with gasoline (even 100% aqueous gasoline-saturated solutions) without exceeding the 5 μg/L MCL for benzene in the pipe water after an 8 hour stagnation period. After 8 hours of stagnation in contact with free product gasoline or soil vapor in close proximity to free product gasoline, the MCL will be exceeded if SBR or NBR gaskets are used. Gasketed PVC pipes (either SBR or NBR Rieber gaskets) should not be used in contact with free product gasoline if conditions of stagnation are likely. DI pipe with NBR Tyton gaskets is an acceptable alternative for such conditions.

PE Service Connections

Small PE service lines should be replaced with impermeable pipe material as soon as it is known that the pipe has been exposed to hydrocarbon contamination, even if the water does not yet have a perceptible odor. Small PE service lines (i.e., 1-inch and ¾-inch) are rapidly permeated by gasoline and gasoline vapors. Concentrations of benzene in groundwater as low as 1 mg/L and 0.6 mg/L will cause the 5 μg/L MCL for benzene to be exceeded in pipe-water after an 8-hour stagnation period for 1-inch and ¾-inch SIDR 9 PE pipes, respectively.

The odor thresholds at 40° C for benzene, toluene, and ethylbenzene are 190 μg/L, 960 μg/L, and 150 ug/L, respectively (Young et al. 1996). At drinking water temperatures, these thresholds would likely be higher. Since most permeation incidents involving service lines are detected by customer complaints of odor, it is highly likely that the MCL has been exceeded by the time of the complaint. This is a particular concern for benzene because the odor threshold is 38 times higher than the MCL.

DI Service Connections

DI service lines joined with SBR Tyton gaskets can be used in groundwater contaminated with gasoline (even saturated with gasoline) without exceeding the 5 μg/L MCL for benzene in the pipe-water after an 8 hour stagnation period. In contact with free product gasoline or soil vapor in close proximity to free product gasoline, the MCL will be exceeded during an 8 hour stagnation period if SBR gaskets are used; however, use of NBR Tyton gaskets will prevent the MCL for benzene from being exceeded. Using NBR gaskets in large service lines in areas that are at risk of hydrocarbon contamination will likely reduce risk of future permeation incidents.

PURE SOLVENTS: BENZENE, TOLUENE, AND TRICHLOROETHYLENE (TCE)

PVC and DI Mains

While PVC pipe is vulnerable to permeation by concentrations of benzene and toluene considerably higher than those found in gasoline or aqueous solutions of gasoline, the gaskets used in water mains are far more permeable and are the limiting factor for engineering decisions in such cases. The permeability of gaskets similarly limits the use of DI pipe in such cases.

Although no experiments were done with pure solvents and gasketed pipe in this project, the guidance given above for the BTEX compounds for concentrations not exceeding those in groundwater saturated with gasoline can be assumed to apply, regardless of the source of the compounds. Those concentrations are: benzene, 72 mg/L; toluene, 85 mg/L; ethylbenzene, 3.4 mg/L; and total xylenes, 19 mg/L.

For higher concentrations of BTEX compounds or for TCE or other compounds, the guidance given in AWWA standards ("consult with the manufacturer regarding permeation.") should be followed.

While PVC pipe would be likely to burst under hydrostatic pressure due to loss of strength before becoming fully rubberized (and, therefore, permeated) by a solvent spill, any PVC pipe that does become permeated without bursting should be replaced because remediation of the pipe is not feasible.

PE Mains

Permeation of hydrocarbons through PE water mains depends on the concentration of contaminant in contact with the pipe, the length of pipe impacted, the formulation of the polymer, and the dimension ratio of the pipe. The tables and equations presented in Chapter 3 can be used to estimate the contaminant concentrations in pipe water for periods of stagnation or flow to formulate site-specific decisions regarding installation of PE mains in contaminated or potentially contaminated areas as well as decisions regarding remediation of installed mains impacted by hydrocarbon contamination.

PE Service Connections

As stated above, since PE will be permeated even for very low hydrocarbon concentrations, small PE service lines should be replaced with impermeable pipe material as soon as it is known that the pipe has been exposed to hydrocarbon contamination, even if the water does not yet have a perceptible odor. For example, concentrations of benzene in groundwater as low as 1 mg/L and 0.6 mg/L will cause the 5 μg/L MCL for benzene to be exceeded in the pipe water after an 8 hour stagnation period for 1-inch and ¾-inch SIDR 9 PE pipes, respectively (See Figure 3.7). Concentrations of toluene in groundwater of 100 mg/L will cause the 1,000 μg/L MCL for toluene to be exceeded in the pipe water after an 8 hour stagnation period for 1-inch SIDR 9 PE pipe (see Figure 3.11).

The odor thresholds at 40° C for benzene and toluene are 190 μg/L and 960 μg/L, respectively (Young 1996). At drinking water temperatures these thresholds would likely be higher. Since most permeation incidents involving service lines are detected by customer complaints of odor, it is highly likely that the MCL has been exceeded by the time of the complaint. This is a particular concern for benzene because the odor threshold is 38 times higher than the MCL.

PVC Service Connections

Free product benzene, toluene, or TCE will rapidly swell (rubberize) PVC pipe, with breakthrough (complete rubberization) occurring in ungasketed 1-inch PVC pipe in 20, 16, and 6.5 days, respectively. Breakthrough for saturated aqueous solutions occurs in 6 to 12 months. No permeation into the potable water occurs until the pipe is completely rubberized (i.e., the moving front reaches the inner wall of the pipe). However, since rubberization seriously degrades the performance characteristics of PVC pipe, pipes attacked in this manner are likely to burst due to hydrostatic pressure before permeation can occur.

A moving front does not form and rubberization does not occur if the level of saturation in groundwater does not exceed 60% for benzene (1,060 mg/L), toluene (300 mg/L), or TCE (656 mg/L).

While PVC pipe would be likely to burst under hydrostatic pressure due to loss of strength before becoming fully rubberized (and, therefore, permeated) by a solvent spill, any PVC pipe that does become permeated without bursting should be replaced because remediation of the pipe is not feasible.

Gasketed Service Connections

Although no experiments were done with pure solvents and gasketed pipe in this project (neither PVC nor DI), the guidance given above for the BTEX compounds for concentrations not exceeding those in groundwater saturated with gasoline can be assumed to apply, regardless of the source of the compounds. Those concentrations are: benzene, 72 mg/L; toluene, 85 mg/L; ethylbenzene, 3.4 mg/L; and total xylenes, 19 mg/L.

For higher concentrations of BTEX compounds or for TCE or other compounds, the guidance given in AWWA standards ("consult with the manufacturer regarding permeation.") should be followed.

COMPATIBILITY OF PVC PIPE WITH HYDROCARBONS

While the gaskets used with PVC pipe are much more susceptible to permeation than PVC pipe itself and are more likely to be the determining factor in engineering decisions related to permeation, questions do arise about the compatibility of PVC pipe materials with hydrocarbons. Those questions are addressed here.

Benzene, Toluene, Ethylbenzene, and Xylenes (BTEX)

PVC pipe is compatible with benzene, toluene, ethylbenzene, and xylenes (BTEX) in groundwater, provided that the sum of the activities of the individual compounds does not exceed 0.40 and no other swelling solvents are present in significant concentrations.

This guidance applies regardless of the source(s) of the BTEX compounds. If gasoline is the only source of contamination, the concentration of BTEX compounds in ground water will not exceed this threshold of compatibility. Gasoline-saturated water has a total BTEX activity of approximately 0.23. No single BTEX compound will permeate PVC pipe if its activity does not exceed 0.40 and no combination of BTEX compounds will permeate PVC pipe if the sum of the individual activities does not exceed 0.40.

The additive compatibility threshold of 0.40 is conservative guidance and allows for margin of safety. Laboratory studies indicated that PVC will not be permeated by BTEX activities ≤ 0.60. Utilities may find the 0.40 activity threshold useful when considering installing new pipes. The 0.60 activity threshold may be useful when making decisions regarding in-service PVC pipes that have become exposed to contamination, but for which no permeation incidents have occurred (i.e., successful uses as defined in Chaper 2).

For this purpose, activity means the concentration of a compound in water expressed as a decimal fraction of its maximum solubility in water. The solubilities of benzene and toluene in water are 1789 mg/L and 530 mg/L for, respectively. Accordingly, the 0.40 activity levels for benzene and toluene are 716 mg/L and 212 mg/L, respectively. These concentrations represent extreme levels of environmental contamination that are not commonly encountered in the field.

Swelling solvents include any compound that will relax the polymer structure of PVC (and many other polymers) and cause permeation of PVC pipe by initiating a moving front. In addition to the BTEX compounds, the swelling solvents that are most often found in contaminated soil and groundwater are the chlorinated solvents such as TCE, PCE and DCE that are used dry cleaning solution, industrial degreasers, and paint strippers; and ketones such as acetone, methylethyl ketone (MEK), and methylisobutyl ketone (MIBK) that are used in laquer thinners and similar products.

Trichloroethylene (TCE)

PVC pipe is compatible with TCE in groundwater at activities up to an activity level of 0.40 (440 mg/L), provided that no other swelling solvents are present in significant concentrations. As explained above for BTEX, this guidance is conservative since laboratory studies indicated that no permeation occurs at activity levels activities ≤ 0.60.

APPENDIX A
CASE HISTORIES

The case histories presented in the following pages were developed from the information in the survey responses and from conversations with water utility officials. Except for one case history that has been publicly presented at a conference, utilities are identified only by size and EPA region. Laboratory data is presented where available. Statements preceded by "presumably" are inferences made by the authors.

Case History A

DI Main w/NBR Gaskets Successfully Installed In Contaminated Soil

Large Municipal Water Utility, U.S. EPA Region 5

A new 10-inch ductile iron water main was being installed as part of a redevelopment of an old industrial area that had been unused and leveled for at least 10 years. Previously, there had been a gas station and a manufacturing plant that made earth-moving equipment in the vicinity of the new water main. During installation of the main, a strong hydrocarbon-like odor was noticed and the clay-like soil had an unusual green appearance. No liquid free product was observed. No laboratory data were obtained for soil or ground water samples. The organic content of the soil is unknown.

Nitrile gaskets, rather than SBR, were used upon recommendation by the supplier of the pipe. The utility also chose to poly-wrap the pipe. While the original plan was to backfill with native soil, the contaminated soil around the new main was removed and the pipe was bedded in sand.

The installation has been successful. The land useage in the area served by the new main is retail/light commercial. There have been no customer complaints regarding odor or taste that might be related to permeation. EPA MCLs for BTEX, chlorinated solvents, or other volatile organic compounds have not been exceeded.

Case History B

Several Successful Uses and One Permeation Incident

Large Municipal Water Utility, U.S. EPA Region 7

Street locations provided by the utility have been deleted.

<u>Mains</u>

An 8-inch PVC main was definitely in contaminated soil near a LUST. There were no taste or odor problems reported with drinking water even though 35 gallons of free product were pumped from a valve box.

A 6-inch CI main was definitely in area of LUST-contaminated soil - no reports of taste/odor.

Lengths of 24-inch CI, 8-inch CI, 6-inch CI, 4-inch CI, and 2-inch CI were all contained within an area of known contamination in the downtown area, referred to as a "no dig" zone. There were no reports of taste or odor complaints.

A 12-inch DI main and a 2 ¼-inch CI main were in an area of known contamination (there was free product in sanitary sewer). There were no reports of taste or odor in the drinking water.

Services

A copper service line was definitely in an area of contamination from a LUST. There were no reports of taste or odor in the drinking water.

Permeation of Service Connection

The gasoline piping to the pumps at a convenience store was leaking, and a PVC water service line passed directly below the area of leakage. There were taste and odor complaints regarding the water inside the convenience store. No laboratory data was obtained. The service line was replaced by the owner, presumably with copper.

Case History C

Successful Use of Gasketed PVC Main in Gasoline-Contaminated Soil

Small Water District, U.S. EPA Region 10

The bedding around an 8-inch, gasketed PVC main failed, allowing the main to sag to the extent that a pipe joint was pulled apart. Hydrocarbon contamination, presumably gasoline from an out-of-business gas station nearby (other side of road), was detected by smell during the excavation. The odor was described as "super bad, strong smelling." Yellow stains were noted on the surface of the PVC. No samples were taken and no laboratory data is available. There had been no customer complaints of bad taste or odor in the section of the system served by that main. SDWA compliance monitoring samples had never been taken from this branch of the system.

A repair was made using a full-circle repair band with a rubber gasket. The utility did not request a special gasket for contaminated soil. The manufacturer stated that the standard gasket used in their products is SBR, but that NBR and other gaskets are available for special applications, including hydrocarbon contaminated soils. Presumably, a SBR gasket was used in this repair.

The repaired main was backfilled with the existing contaminated soil and returned to service. There were no remediation efforts. There continued to be no customer complaints after the repair. No water quality monitoring data was available for this branch of the system.

The pipe was Class 155 8-inch bell end gasketed blue PVC, originally installed in 1978. The manufacturer is unknown. The gaskets are double o-ring of unknown type (but presumably

SBR). The soil is of uncertain type and organic content, otherwise described as "coastal terrace material." It is not known when the gasoline contamination began.

Case History C1

Successful Use of PB Service Line in Gasoline-Contaminated Area

Small Water District, U.S. EPA Region 10

This service line was connected to the main described in Case History C.

An out-of-business gas station was located across the road from the site of the water main failure described in Case History C. While the gas station had been closed for some time, an upstairs apartment in the same building was occupied at the time of the failure of the main. A modular home had been built behind the gas station building and was also occupied at that time. Both of these residences were served by polybutylene (PB) service lines (size not specified).

The soil where the main was tapped for this service line was contaminated, presumably with gasoline from the closed gas station. The odor of the soil excavated during the water main repair was described as "super bad, strong smelling." It is not known what fraction of the length of the PB service line was in contaminated soil. The soil is of uncertain type and organic content, otherwise described as "coastal terrace material." The main was originally installed in 1978. It is not known when the gasoline contamination began. There had been no complaints of bad taste or odor from the residents.

The gas station building has since been demolished. The modular home continues to be occupied without customer complaints of bad taste or odor. No soil samples or water samples have been taken at this residence.

Case History D

Successful Uses of Riveted Steel and Cased, Wrapped DI/NBR Water Transmission Mains In Soil Contaminated By Petroleum-Hydrocarbons

Akron, Ohio, Public Utilities Bureau

Based on a paper entitled "Transmission Main Installation Through Contaminated Soil" presented by Gregg Loesch, P.E. at the 2002 ASCE Pipelines Conference, Cleveland, OH, and on communications with the author.

Old riveted steel 36-inch and 48-inch water transmission mains were being replaced by 48-inch prestressed concrete mains when a 525 ft long area of significant petroleum product contamination was encountered when the trench depth was increased to cross under an existing storm culvert, petroleum pipeline, and railroad. Workers noticed a "volatile gasoline smell" at a trench depth of 12 ft. As the excavation progressed, the smell intensified and a liquid, presumably a petroleum product, appeared at the bottom of the trench. The excavation was halted and the replacement in this area was delayed for a number of years while this part of the project was re-engineered. A soil sample associated with the free product pocket contained 162 mg/kg total BTEX. There had been no complaints or laboratory data exceeding U.S. EPA MCLs that were attributable to this area of contamination.

The contaminated area was eventually crossed using 1000 ft of specially designed water mains with several protective features, keeping a minimum depth of 3 ft of undisturbed soil between the contamination and the bottom of the trench. The mains were constructed in the following manner:

- 36-inch Class 53 DI pipe with NBR (nitrile) gaskets.
- Wrapped in an 8 oz. impervious geotextile material (Petrogard VI, MPC Containment Systems, Chicago, IL)
- Encased in low strength mortar backfill to 12 inches above the top of the pipe.
- Surrounded by a 16 oz. trench liner (Petrogard X, MPC Containment Systems, Chicago, IL) extending to within 18 inches of the finished grade.
- Covered by a granular backfill material.

There have been no customer complaints or laboratory data exceeding U.S. EPA MCLs that could be attributed to this area of contamination since specially designed transmission mains were installed.

Case History E

Permeation of a PE Residential Service Line

Large Municipal Water Utility, U.S. EPA Region 6

Customers complained of bad taste. An extensive investigation traced the contamination to a LUST at a convenience store. PE service lines were replaced with copper in two residences that were 300 to 400 feet away from the LUST. There are soil and groundwater plume maps for total BTEX and TPH associated with this incident. The ¾-inch PE service lines were in native soil described as sandy clay with gravel and of low (<1%) organic content. The time of exposure prior to the complaint was unknown. The data for the monitoring well closest to the residence in the more highly contaminated part of the site is given below.

Matrix	Analyte	Concentration	Units
Soil	Total BTEX	7.7	mg/kg
Soil	TPH	24	mg/kg
GW	Total BTEX	57.7	mg/L
GW	TPH	45	mg/L

Case History F

Successful Use of DI/SBR Main in PCE-Contaminated Soil
Permeation of DI/SBR Service Line

Large Municipal Water Utility, U.S. EPA Region 7

Customers complained of improper disposal of dry cleaning solvents. The soil, ground water, and potable water were found to contain PCE. The utility described the site as an "unconfirmed but suspected" disposal site. Both the main (8-inch) and the service line were DI pipe with SBR gaskets, bedded in native clay and loam soils. The service line was relocated away from the contaminated area. The flow in the main dilutes the PCE concentration below the MCL of 5 μg/L. The data associated with this incident are given below.

Matrix	Analyte	Concentration	Units
Soil	PCE	0.036 - 3.600	mg/kg
GW	PCE	0.90 - 15.4	mg/L
PW	PCE	ND - 0.00675	mg/L

Case History G

Successful Use of DI/FKM Main and Service Connection
In Soil Contaminated By Chlorinated Solvents

Medium Municipal Water Utility, U.S. EPA Region 7

A 16-inch DI main with FKM and a 6-inch DI service line, also with FKM gaskets, were bedded in native sandy loam that was impacted by chlorinated solvents from a disposal site. The source of the solvents was undetermined. Total VOC in the groundwater was 0.715 mg/L. There were no customer complaints, and no chlorinated solvents were detected in the potable water.

Case History H

Permeation of PVC Main by Chlorinated Solvents

Medium County Water Utility, U.S. EPA Region 9

Customers served by a 2-inch PVC main (Schedule 120) complained of bad taste and odor. The main was found to be in soil that had been saturated by a spill of chlorinated solvents.

The exposure time prior to the complaint was unknown. The type of jointing was not reported. The main was replaced with metal pipe, possibly copper. The potable water, sampled at the time of the complaint, contained 5.8 µg/L 1,2-dichlorobenzene and 2 µg/L 1,4-dichlorobenzene.

Case History I

Successful Use of PVC/FKM Main in Soil Contaminated with Chlorinated Solvents

Small Water and Sewer District, U.S. EPA Region 8

Private residential wells were found to be contaminated by chlorinated solvents from a Superfund site. The utility extended potable water service to the area using 10-inch and 8-inch PVC mains with FKM gaskets, and copper service connections. Groundwater data from several monitoring wells associated with these mains is presented below.

Analyte	Concentrations	Units
PCE	250, 262, 96, 952	µg/L
TCE	73, 119, 13, 133	µg/L
cis-DCE	241, 266, -, 386	µg/L
VC	17, 13, 12, 34	µg/L

This was reported case of a utility using nothing other than DI pipe in area of known contamination.

Case History J

Successful Use of CI/SBR MJ Main in Contaminated Soil
Permeation of DI/SBR Service Line
Successful Use of DI/FKM Service Line

Large Municipal Water Utility, U.S. EPA Region 7

A 12-inch CI main with SBR-gasketed mechanical joints was found to be in soil contaminated by contaminants from a LUST. A 6-inch DI/SBR service line was also within this contaminated area. The pipes were bedded in native loam and clay soils. The time of exposure prior to discovery was unknown. The utility determined that contaminants entered the main and service connection through the SBR gaskets. There were no customer complaints. The service line was abandoned in place and replaced with 6-inch DI/FKM. The laboratory data associated with this case are presented below.

Matrix*	Analyte	Concentration	Units
Soil	2-methylnaphthalene	0.37	mg/kg
GW	acetone	0.720 - 0.790	mg/L
GW	Total xylene	0.055 - 0.180	mg/L
GW	naphthalene	0.059	mg/L
GW	2-methylnaphthalene	0.089	mg/L
GW	phenanthrene	0.045	mg/L
GW	Total extractable hyd.	24.8	mg/L
GW	diesel	10.1	mg/L
GW	gasoline	12.8	mg/L
GW	motor oil	1.9	mg/L
GW	benzene	0.0258	mg/L
GW	toluene	0.0144	mg/L
GW	ethylbenzene	0.0771	mg/L
PW	toluene	0.0001-0.0002	mg/L
PW	ethylbenzene	0.0001	mg/L
PW	m/p-xylene	0.0002	mg/L
PW	o-xylene	0.0001	mg/L
PW	benzene	ND	mg/L

* GW = groundwater   PW = potable water

Case History K

Many Permeations of Polybutylene Service Lines

Large Municipal Water District, U.S. EPA Region 9

The utility reported 11 incidents involving permeation of PB service lines. In all cases, the PB pipes were intact and the customers had complained of odor in the potable water. Five incidents were attributed to gasoline, and one incident each was attributed to acetone, kerosene, toluene/xylene, and a "petroleum product". Two incidents were attributed to unknowns. In all cases, the PB service lines were replaced with copper. In the incident attributed to acetone, the acetone concentration was 44.7 g/kg in the soil and 16 mg/L in the potable water.

Case History L

Three Failures of PVC/SBR Mains

Small Community Water and Sewer Association, U.S. EPA Region 5

In the following incidents, the pipes were PVC SDR 26 (ASTM D 2241) joined with SBR gaskets and were bedded in native clay soil. Times of exposure resulting in failure were unknown. The joints were not damaged, but a section of each pipe was blistered and leaking, and very brittle when replaced. Damaged sections of pipe were cut out and replaced using repair couplings with SBR gaskets. Contaminated soil was replaced with clean soil of the same type from the surrounding area. The repairs continued to be effective until the date of this report (11 to 16 years). There were no laboratory data available for these incidents.

Incident L1

Customers complained of taste and odor in the potable water and a leak in a 2-inch PVC main on private property. The utility determined that the incident resulted from a spill or disposal of an unknown contaminant on private property. The main was thoroughly flushed after repair.

Incident L2

A leaking main was reported adjacent to garage that had burned. The type of contaminant was unknown, but the garage was known to contain exotic drag racing fuels.

Incident L3

A leaking main was reported adjacent to a petroleum storage facility. The utility determined that the contaminants were gasoline and diesel fuel.

These incidents account for three of the four permeations of PVC reported in the water utility survey. While they are recorded as permeation incidents in the survey analysis, they may actually have been infiltrations since leaks were involved. These are the only incidents for which PVC pipe was reported to be blistered, leaking, and brittle.

Case History M

Many Permeations of Polybutylene Service Lines

Large Municipal Utility District, U.S. EPA Region 9

The utility reported 23 incidents involving permeation (without physical failure) of 1-inch PB service lines between 1990 and 2004. No laboratory data was available.

The nature of the customer complaints associated with these incidents was reported as follows:

Taste and odor	9
Odor only	12
Taste only	1
No record	1

The identities of the contaminants were reported as follows:

No record	16

Gasoline	5
Chlorinated Solvents	2

Of the five incidents attributed to gasoline, four resulted in odor (only) complaints and one resulted in a taste (only) complaint. Both incidents attributed to chlorinated solvents resulted in odor (only) complaints.

In all cases, the service lines involved were replaced with copper.

Case History N

Permeation by Gasoline of SBR Gaskets In Mechanically Jointed Cast Iron Pipe

Medium Municipal Water Utility in U.S. EPA Region 5

A 16-inch CI main was excavated for repairs due to a leak and the clay soil in the excavation smelled of gasoline. It was observed that the mechanical joint bolts and gaskets were "eaten up." The bolts and gaskets were replaced and the contaminated soil in the excavated area was replaced with sand. The gaskets were replaced with "standard" gaskets, presumably SBR. No laboratory data was available. Other mechanical joints in the area were investigated and signs of external corrosion on bolts and exposed areas of gaskets were observed.

Many years before this incident, two gas stations were located adjacent to the main in the vicinity of the failures. Although the gas stations no longer existed and their USTs had been removed, it was assumed that the source of the contamination was leaking USTs from those stations.

All of the bolts in that section of main were replaced in connection with a street reconstruction two years after the incident. The gaskets in two joints were replaced at that time with "standard" gaskets, presumably SBR. It was noted that joints appeared to be in progressively worse condition as proximity to the UST sites increased.

Case History O

Permeation of an Asbestos-Cement Main by Gasoline

Large Utility in Western Canada

The utility became aware of leaking underground storage tank (LUST). The water in a 350 mm (14-inch) asbestos-cement (AC) main with SBR gaskets was checked and found to be contaminated (no laboratory data were available). The soil type was clay and the main was bedded in sand.

The LUST was removed and all contaminated soil on the property and road right of way was removed and replaced with new soil. The main and services in the contaminated area were replaced with new materials of the same size and type as the originals. The repair has been considered successful for at least eight years.

Case History P

Permeation of a PVC Distribution System by Gasoline

Large County Water Utility in U.S. EPA Region 4

BTEX contamination was noted in the privately owned water distribution system of a condominium development. The development purchased water from the county water system through a master meter and distributed it within the development using an 8-inch PVC main (presumably with SBR gaskets) and PVC service laterals. A convenience store adjacent to the development had a leaking underground storage tank (LUST). County water officials believed that the contamination entered the distribution system through the service laterals, presumably by permeation of the gaskets.

The remediation was settled between the private parties owning the convenience store and the condominium development. The LUST was replaced and the contaminated soil was replaced with new soil. The distribution system was left in service without replacement.

Case History Q

A Proactive Approach to Permeation

Medium Municipal Utility in U.S. EPA Region 8

The utility has LUST sites, PCE contaminated areas resulting from dry cleaning, and a site contaminated by an old oil refinery in its service area. The utility has used “a proactive methodology for water pipe materials in known or suspected contaminated areas since the late 1980s” and has never had a known permeation incident.

While the utility has used PVC mains almost exclusively because of corrosive soils, DI pipe with resistant gaskets are required in known or suspected contaminated areas. Nitrile (NBR) or FKM gaskets are required depending on the severity of contamination. Only flared copper pipe is allowed for service connections in known or suspected contamination. The utility performs quarterly monitoring of the distribution system in known contaminated areas to assure that contamination is not entering through nonresistant gaskets installed before the proactive policies were instituted.

The utility reported the existence of 300 feet of PVC main in an area where the groundwater is known to be contaminated with PCE. No problems had been noted.

Three examples of successful uses resulting from the proactive policies were reported (all mains were bedded in sand):

A 12-inch DI main with FKM gaskets was installed in sandy soil in a RCRA brownfields project area with contamination from gasoline, diesel fuel, fuel oil, and chlorinated solvents. Service connections were also installed, presumably using flared copper as above.

A DI main (size not reported) with nitrile gaskets was installed in sandy soil that had been contaminated with chlorinated solvents (PCE) from dry cleaning. Service connections were also installed, presumably using flared copper as above.

A DI main (size not reported) with nitrile gaskets was installed in clay soil that had been contaminated with gasoline from a LUST. Service connections were also installed, presumably using flared copper as above.

No laboratory data were reported.

Case History R

Contamination Discovered During Planned Replacement

Small Municipal Utility in Central Canada

The utility did a planned replacement of an old 8-inch cast iron main that had machined joints with red lead pipe dope. Upon excavation, it was discovered that the soil was contaminated with gasoline from LUSTs on an adjacent property. The main was replaced with 10-inch PVC with NBR gaskets. The LUSTs were subsequently removed. No further remediation was reported. The PVC main had been in service for 9 years at the time of the report, presumably without incident.

Case History S

Diesel Fuel in a Meter Box

Large Municipal Utility in U.S. EPA Region 7

The utility received three taste and odor complaints from a multi-family residence. Upon investigation, it was discovered that diesel fuel had entered the meter box as a result of a spill which had occurred two days earlier. The service line was ¾-inch PE with a butadiene rubber (BR) gasket in the meter box. The soil type was loam. The utility concluded that the contaminants had permeated through the gasket. The authors presume that this was actually an SBR gasket.

The fuel was cleaned from the meter box and the contaminated soil was removed and replaced. The service line from the main to the residence was replaced with copper.

Authors' note: Work reported elsewhere in this report showed that free product gasoline takes approximately one week before BTEX can be detected in the water inside a 1-inch PE pipe. Diesel fuel would likely require a longer time for a breakthrough sufficient to result in customer complaints, because the concentration of BTEX compounds in diesel fuel is lower than in gasoline. Other researchers have shown that SBR gaskets are considerably more permeable than PE. Consistent with the utility's conclusion, it is likely that the initial contamination that resulted in the customer complaints did occur through the gasket. A prolonged period of contact would also have resulted in permeation through any exposed PE pipe.

Case History T

Solvents in a Meter Box

Large County Utility in U.S. EPA Region 4

The utility received an odor complaint from a customer. Upon investigation, it was discovered that asphalt solvents had entered the meter box as a result of a spill that had occurred five days earlier. The concentration of n-butylbenzene in the soil was 0.34 mg/kg. There were no detectable contaminants in the potable water in the residence. The service line was ¾-inch copper with an EPDM gasket in the meter box. The soil was a low-organic sandy soil and the pipe bedding was sand. Presumably, the odor-causing contaminants permeated through the EPDM gasket.

The contaminated soil was removed and replaced. Presumably, the meter box was also cleaned. Whether or not the gasket was replaced was not reported.

Case History U

Service Line Contaminated by Fuel Oil From a LUST

Large Municipal Utility in U.S. EPA Region 10

The utility received a customer complaint of taste and odor. Upon investigation, it was determined that BTEX from fuel oil had permeated the ¾-inch PB service line as a result of a LUST. The estimated time of exposure prior to the complaint was approximately 36 months. Total BTEX in the potable water in the residence was 15 μg/L. The service line was replaced with copper.

The utility stated that other incidents had occurred within their distribution system but that information for those was incomplete.

Authors' note: Work reported elsewhere in this report showed that benzene accounts for somewhat more than half of the total BTEX permeating PE pipe at low concentrations of gasoline in water (toluene accounts for the remainder). This is likely true for PB pipe as well. Given that the total BTEX in the potable water was 15 μg/L at the time of the complaint, it is probable that the EPA MCL for benzene of 5 μg/L had been exceeded at that time. The authors have observed in studies involving PE pipe exposed to gasoline that the concentration of benzene exceeds 5 µg/L before an odor is noticeable. The odor threshold for benzene and toluene are 190 µg/L and 960 µg/L, respectively (see Chapter 9).

Case History V

Curb Box Contaminated by Customer

Large Municipal Utility in U.S. EPA Region 6

The utility received a complaint from a residential customer of bad taste and odor.. Upon investigation, water at the outdoor tap at the residence was found to taste bitter and the meter box was found to smell strongly of gasoline. The customer stated that he had changed the fuel pump in a vehicle three weeks earlier, that fuel had leaked from the vehicle during that work, and that he had washed the driveway with a garden hose. The meter box was down-slope from the driveway. The customer also indicated that he had been ill for the past two weeks and that he attributed that illness to the drinking water. Utility personnel took a water sample from the residence for laboratory analysis, removed soil which smelled like gasoline from the meter box, and allowed the meter box to air out for several days.

Six days later, the utility received the laboratory report indicating that the water was contaminated with hydrocarbon substances resembling gasoline and cleaning solvents. Additional samples were collected at the residence, the corporation stop for the residence, and at two nearby residences. The service was excavated from the 6-inch PVC main to the meter box, and the utility-owned service line was determined to be PE, as were the pigtail and the customer-owned service line.

The second set of samples was analyzed immediately and the results indicated that contamination was still present at the residence but at lower levels that in the initial sample. No contamination was detected at the other residences or in the main. The problem was shown to be isolated to the complaining residence. Service to the residence was kept disconnected pending further investigation.

	Day 0	Day 6, vial 1	Day 6, vial 2
Compound	µg/L	µg/L	µg/L
Benzene	76	19	83
Toluene	59	58	280
2-Chlorotoluene	266	86	133
4-Chlorotoluene	131	42	68
Chlorobenzene	0.4	ND	ND
1,2-Dichlorobenzene	12.5	ND	ND
1,2,4-Trichlorobenzene	3	ND	ND
1,2,3-Trichlorobenzene	0.4	ND	ND
Ethylbenzene	29.4	ND	36
o-Xylene	0.4	ND	6.3
Isopropyl benzene	2.1	ND	1.5
tert-Butylbenzene	0.3	ND	ND

sec-Butylbenzene	0.4	ND	ND
n-Butylbenzene	15.6	ND	ND

On the following day (Day 7), the utility owned service line was replaced with copper. The customer's service was restored on Day 8 upon the customer's request. The customer was advised about methods for cleanup of his property.

Samples were collected on Day 15 at the complaining residence and at the residence next door. No contaminants were detected at either location. The customer was advised that there was still a possibility that VOCs in the soil could permeate the customer-owned plastic service line. The customer requested copies of the lab reports. The customer was responsible for the connection from the meter to the house and the utility is uncertain whether the pigtail and the service line were ever replaced with copper. There were no further complaints from this customer (more than 4 years since the incident).

Case History W

Gasoline Spill Permeates Service Line

Large Municipal Utility in U.S. EPA Region 3

The utility received a customer report of bad odor in the drinking water. Upon investigation, the utility determined that a gasoline spill had permeated the PB service line and entered the water. The pipe was in a sandy, low organic soil and was bedded in sand.

The service line was replaced with copper. The utility observed that the gasoline had turned the PB pipe "into spaghetti."

Case History X

Leaking Gas Tank Permeates Service Line

Large Municipal Utility in U.S. EPA Region 4

The utility received a report from a residential customer of bad odor in the drinking water. Upon investigation, it was determined that the contamination was probably from the leaking fuel tank of a vehicle in the driveway. The PE service line paralleled the driveway and appeared to be permeated by hydrocarbons. The service line was replaced with copper.

Case History Y

Successful Use of CI Pipe with Leaded Joints In a Superfund Site

Medium Municipal Water Utility, U.S.E.P.A. Region 7

Two similar cases were reported, four years apart, involving exposure of CI pipe with leaded joints exposed to chlorinated solvents from a Superfund site. Mains and services, from 4-inch to 12-inch in size, were installed prior to 1950 in sandy loam soil.

The first case involved dry cleaning solvents and a bulk storage facility. PCE concentrations were 1,560 mg/kg in the soil and 0.440 mg/L in the groundwater. No chlorinated compounds were detected in the potable water.

The second case involved dry cleaning solvents resulting in a PCE concentration of 59.0 mg/kg in the soil. No chlorinated compounds were detected in the potable water.

Case History Z

Successful Use of DI Pipe with SBR Gaskets Near a Bulk Storage Facility

Medium Municipal Water Utility, U.S.E.P.A. Region 7

Class 52 DI mains and services, from 6-inch to 16-inch, were installed in sandy loam that was contaminated by leakage of chlorinated solvents from a bulk storage facility, resulting in a total VOC concentration in the groundwater of 0.014 mg/L. No chlorinated compounds were detected in the potable water.

Case History AA

Successful Use of PVC Main Impacted by a LUST

Large Municipal Water District, U.S.E.P.A. Region 9

The utility received a report of a LUST and discovered that leaking gasoline was in contact with a PVC main. Presumably, the main was joined with SBR gaskets. Laboratory analysis of samples taken a few days after the utility became aware of the situation indicated 0.05 mg/L total purgeable hydrocarbons (TPH) in the potable water. The utility did not consider this to be a failure because the THP was at the laboratory detection limit and, presumably, no EPA MCLs had been exceeded. No additional information was reported.

REFERENCES

Astarita, G., and S. Joshi. 1978. Sample-Dimension Effects in the Sorption of Solvents in Polymers - A Mathematical Model. *Jour. Membrane Science*, 4(1):165-182.

Barnes, S.E., E.C. Brown, M.G. Sibley, H.G.M. Edwards, and P.D. Coates. 2005. Vibrational Spectroscopic and Ultrasound Analysis for the In-process Monitoring of Poly(ethylene vinyl acetate) Copolymer Composition During Melt Extrusion. *Analyst,* 130: 286-292.

Barrer, R.M. 1951. *Diffusion in and through Solids*. Cambridge, U.K.: University Press.

Berens, A.R., and H.B. Hopfenberg. 1982. Diffusion of Organic Vapors at Low Concentrations in Glassy PVC, Polystyrene, and PMMA. *Jour. Membrane Science*, 10(2-3):283-303.

Berens, A.R. 1985. Prediction of Organic Chemical Permeation through PVC Pipe. *Jour. AWWA*, 77(11):57-64.

Berens, A.R. 1997. Predicting the Migration of Endocrine Disruptors from Rigid Plastics. *Polymer Engineering and Science*, 37(2):391-395.

Bird, W. 2006. Private Communication. S&B Technical Products, 1300 East Berry St., Fort Worth, TX 76119.

Bohacs, G., Z. Ovadi, and A. Salgo. 1998. Prediction of Gasoline Properties with Near Infrared Spectroscopy. *Jour. Near Infrared Spectrosopy*, 6:341-348.

Burn, S., P. Davis, T. Schiller, B. Tiganis, G. Tjandraatmadja, M. Cardy, S. Gould, P. Sadler, and A. J. Whittle. 2005. Long-Term Performance Prediction For PVC Pipes. Denver, Colo: Awwa RF and AWWA.

Centner, V., J. Verdu-Andres, B. Walczak, D. Jouan-Rimbaud, F. Despagne, L. Pasti, R. Poppi, D. Massart, and O.E. Noord. 2000. Comparison of Multivariate Calibration Techniques Applied to Experimental Data Sets. *Jour. of Applied Spectroscopy,* 54(4): 608-623.

Chevron Phillips Chemical Company. 2003. Fluid Flow. In *The Performance Pipe: Engineering Manual*, Book2: System Design, Chapter 4. Plano, TX: Chevron Phillips Chemical Company LP.

Chiou, C.T., and L.J. Peters. 1981. Soil-Water Equilibrium for Nonionic Organic Compounds. *Science*, 213(4508):683-684.

Choquette, S.J., J.C. Travis, Z. Changjiang, and D.L. Duewer. 2002. Wave Number Standards for Near Infrared Spectrometry. In *Handbook of Vibrational Spectroscopy.* Edited by John Chalmers and Peter R. Griffiths. UK: Wiley.

Chung, H., H. Lee, and C.H. Jun. 2001. Determination of Research Octane Number Using NIR Spectral Data and Ridge Regression. *Jour. Bulletin of Korean Chemistry Society*, 22(1):37-42.

Cline, P.V., J.J Delfino, and P.S.C. Rao. 1991. Partitioning of Aromatic Constituents into Water from Gasoline and Other Complex Mixtures. *Environmental Science Technology*, 25(5):914-920.

Corbett, B.G. 2006. Private Communication. S&B Technical Products, 1300 East Berry St., Fort Worth, TX 76119.

Crank, J. and G.S. Park. 1968. *Diffusion in Polymers*. New York: Academic Press.

Crank, J. 1975. *The Mathematics of Diffusion*. Oxford, UK: Clarendon Press.

Ductile Iron Pipe Research Association 2006, Hydraulic Analysis of Ductile Iron Pipe, Birmingham, AL.

Dryden, G.M. 2003. Near Infrared Reflectance Spectroscopy: Applications in Deer Nutrition. A Report for the Rural Industries Research and Development Corporation (Publication No W03/007, Project No UQ 109A) [Online]. Available: <http://www.rirdc.gov.au/reports/DEE/w03-007.pdf>. [cited June 20, 2005]

Fick, A., 1855. *Poggendorff's Annel. Physik.* 94:59.

Fischer, D., and K.-J. Eichhorn. 1998. Process Analysis of Polymers by NIR Spectroscopy. *Analysis Magazine,* 26(4): M58-M61.

Fischer, D., and E. Pigorsch. 2000. New Developments in Process Control by Spectroscopic Methods in the Polymer and Plastics Industry-Near Infrared Miniature Spectrometer and High Temperature and Pressure Near Infrared and Raman Probes. Paper for the Third Annual UNESCO School IUPAC Conference on Macromolecules and Materials Science, Matieland, South Africa., April 10- April 12, 2000.

Fourier, J.B. 1822. *Theorie Analytique de la Chaleur*, English translation by A. Freeman. New York: Dover Publication.

Glaza, E.C., and J.K. Park. 1992. Permeation of Organic Contaminants through Gasketed Pipe Joints. *Jour. AWWA*, 84(7):92-100.

Guo, C.J., K.D. De, and B. Harrison. 1992. Effect of Molecular Structure on Diffusion of Organic Solvents in Rubbers. *Chemical Engineering Science*, 47(7):1525-1532.

Harogoppad, S.B., and T.M. Aminabhavi. 1992. Diffusion and Sorption of Organic Liquids through Polymer Membranes. VIII. Elastomers versus Monocyclic Aromatic Liquids. *Jour. Applied Polymer Science,* 46(4):725-732.

Holsen, T.M., J.K Park, D. Jenkins, and R.E. Selleck. 1991a. Contamination of Potable Water by Permeation of Plastic Pipe. *Jour. AWWA*, 83(8):53-56.

Holsen, T.M., J.K. Park, L. Bontoux, D. Jenkins, and R.E. Selleck. 1991b. The Effect of Soils on the Permeation of Plastic Pipes by Organic Chemicals. *Jour. AWWA*, 83(11):85-91.

Hopfenberg, H.B. 1978. The Effect of Film Thickness and Sample History on the Parameters Describing Transport in Glassy Polymers. *Jour. Membrane Science*, 3(2):215-230.

Hruschka, W.R. 1987. Data Analysis: Wavelength Selection Methods. In *Near–Infrared Technology in Agricultural and Food Industries*. Edited by P. Williams and K. Norris. St. Paul, Minn.: American Association of Cereal Chemists.

Hui, C.Y., K.C. Wu, R.C. Lasky, and E.J. Kramer. 1987a. Case-II Diffusion in Polymers. I. Transient Swelling. *Jour. Applied Physics*, 61(11):5129-5136.

Hui, C.Y., K.C. Wu, R.C. Lasky, and E.J. Kramer. 1987b. Case-II Diffusion in Polymers. II. Steady-State Front Motion. *Jour. Applied Physics*, 61(11):5137-5149.

Ingels, L. 2004. Private Communication. American Gas Association, 400 N. Capitol St., NW, Suite 450, Washington, DC, 20001

Joo, J.C., J.Y. Kim and K. Nam. 2004. Mass Transfer of Organic Compounds in Dilute Aqueous Solutions into High Density Polyethylene Geomembranes. *Jour. of Environmental Engineering*, 130(2):175-183.

Jutras, E.M., C.M. Smart, R. Rupert, I.L. Pepper and R.M. Miller. 1997. Field-Scale Biofiltration of Gasoline Vapors Extracted from Beneath a Leaking Underground Storage Tank. *Biodegradation*, 8(1):31-42.

Khinnavar, R.S., and T.M. Aminabhavi. 1992. Resistance of Barrier Elastomers to Hazardous Organic Liquids. *Jour. Applied. Polymer Science*, 45(4):1107-1125.

Klute, A. 1986. *Methods of Soil Analysis: Part I: Physical and Mineralogical Methods.* Monograph Number 9. Madison, WI: American Society of Agronomy.

Lauder, T.J. 2004. Private Communication. PSE & G Gas Delivery, 80 Park Plaza, Newark, NJ 07101.

Lee, P.I., and C.J. Kim. 1992. Effect of Geometry on Solvent Front Penetration in Glassy Polymers. *Jour. Membrane Science*, 65(1-2):77-92.

Matthews, G. 1996. *PVC: Production, Properties and Uses*. London, U.K: Institute of Materials.

Menting, L.C., B. Hoogstad, and H.A.C. Thijssen. 1970. Diffusion Coefficients of Water and Organic Volatiles in Carbohydrate-Water Systems, *Jour. of Food Technology*, 5(2):111-126.

National Institute of Standards & Technology. 1988. *Lead in Reference Fuel.* Standard Reference Matherial 2712. Gaitbersburg, MD.: NIST.

National Bureau of Standards. 1986. *Alcohols in Reference Fuel.* Standard Reference Material 1829. Gaitbersburg, MD.: NBS (now NIST).

Naylor, T. D., 1989. Permeation Properties. In *Comprehensive Polymer Science*. Edited by G. Allen and J.C. Bevington. Toronto, Canada: Pergamon Press.

Olson, A.J., D. Goodman, and J.P. Pfau. 1987. Evaluation of Permeation of Organic Solvents through PVC, Asbestos/Cement, and Ductile Iron Pipes. *Jour. of Vinyl and Additive Technology*, 9(3): 114-118.

Ozdemir, D. 2005. Determination of Octane Number of Gasoline Using Near Infrared Spectroscopy and Genetic Multivariate Calibration Methods. *Jour. Petroleum Science and Technology*, 23:1139-1152.

Page, A.L. 1986. *Methods of Soil Analysis: Part II: Chemical and Microbiological Properties.* Monograph Number 9. Madison, WI: American Society of Agronomy.

Park, J.K., L. Bontoux, T.M. Holsen, D. Jenkins, and R.E. Selleck. 1991. Permeation of Polybutylene Pipe and Gasket Material by Organic Chemicals. *Jour. AWWA*, 83(10):71-78.

Park, J.K., and L. Bontoux. 1993. Thermodynamic Modeling of the Sorption of Organic Chemicals in Thermoplastics and Elastomers. *Jour. Applied Polymer Science*, 47(5):771-780.

Park, J.K., J.P. Sakti and J.A. Hoopes. 1996. Transport of Aqueous Organic Compounds in Thermoplastic Geomembranes II: Mass Flux Estimates and Practical Implications. *Jour. Environmental Engineering*, 122(9):807-813.

Rahman, S., and N. Alchin. 2005. Riebert Joint System for PVC Pipe: Synopsis of a Locked-In Elastomeric Gasket. In *Proc. Pipe 2005 Conference*. Wagga, NSW, Australia.

Reckhorn, S.B.F., L.V. Zuquette, and P. Grathwohl. 2001. Experimental Investigations of Oxygenated Gasoline Dissolution. *Jour. of Environmental Engineering*, 127(3): 208-216.

Reich, G. 2005. Near-infrared and imaging: Basic principles and pharmaceutical applications. *Advanced drug delivery reviews,* 57:1109-1143

Rittmann B.E., and P.L. McCarty. 2001. *Environmental Biotechnology: Principles and Applications.* New York: McGraw-Hill.

Rohe, T., W. Becker, A. Krey, H. Nägele, S. Kölle, and N. Eisenreich. 1998. In-line Monitoring of Polymer Extrusion Processes by NIR Spectroscopy. *Jour. Near Infrared Spectroscopy*, 6:325–332.

Saleem, M., A.A. Asfour, K.D. De, and B. Harison. 1989. Diffusion of Organic Penetrant through Low Density Polyethylene (LDPE) Films: Effect of Size and Shape of the Penetrant Molecules. *Jour. Applied Polymer Science*, 37(3):617–625.

Sangam, H.P., and R. K. Rowe. 2001. Migration of Dilute Aqueous Organic Pollutants through HDPE Geomembranes. *Geotextiles and Geomembranes*, 19(6):329-357.

Satterfield, C.N. and C.S. Cheng. 1971. Single-Component Diffusion of Selected Organic Liquids in Type Y Zeolites. *Jour. American Institute of Chemical Engineers*, 18(4):720–723.

Schwarzenbach R.P., P.M. Gschwend, and D.M. Imboden. 1993. *Environmental Organic Chemistry*. New York: Wiley and Sons.

Seo Y.C., and N.J. Kim. 1998. Permeation of Organic Chemicals Through Gasketed Cast Iron Pipe. *J. Korean Environ. Sci. Society*. 8:45-50.

Thermo electron. 2005. Spectroscopic Solutions 2005. Material from the one-day seminar carried out by Thermo Electron in St. Paul, Minneapolis.

Thomas, N.L. and A.H. Windle. 1982. Theory of Case II diffusion. *Polymer*, 23(4):529-542.

Thompson C. and D. Jenkins. 1987. *Review of Water Industry Plastic Pipe Practices, Report to Awwa Research Foundation,* Order # 90516; Denver, CO: AwwaRF and AWWA.

Uni-Bell. 2001. *Handbook of PVC Pipe Design and Construction*. Dallas, TX.: Uni-Bell PVC Pipe Association.

US EPA. (U.S. Environmental Protection Agency) 2001. Permeation and Leaching: Technical Information Review [Online]. Available from Internet <http://www.epa.gov/safewater/disinfection/tcr/pdfs/whitepaper_tcr_permation-leaching.pdf> [cited August 2, 2006]

Vasile, C., and R.B. Seymour. 1993. *Handbook of Polyolefins: Synthesis and Properties*, New York: Marcel Dekker, Inc.

Volland, W. 1999. Organic Compound Identification Using Infrared Spectroscopy. [Online] Available: <http://www.800mainstreet.com/irsp/eir.html> [cited: May 10, 2006]

Vonk, M.W., and G. Veenendaal. 1984. Permeation of Methyl Bromide and Other Chemicals through PE and PVC Pipes for Drinking Water Supply. *Water Supply,* 2:61-69.

Vonk, M.W. 1985. Permeation of Organic Compounds through Pipe Materials. Publication No. 85. Neuwegein, Netherlands: KIWA.

Vonk, M.W., 1986. Permeation of Organic Soil Contaminants through Polyethylene, Polyvinylchloride, Asbestos-Cement and Concrete Water Pipes. In *Proc. of Water Quality Conference.*

Williams, P., and K. Norris. 2001. *Near-Infrared technology in the agricultural and food industries*, Williams, J. and K. Norris (Eds.), St. Paul, MI: American Association of Cereal Chemists, Inc.

Xiao, J. and J. Wei. 1992. Diffusion Mechanism of Hydrocarbons in Zeolites. *Chemical Engineering Science*, 47(5):1123-1141.

Young, W. F., H. Horth, R. Crane, T. Ogden, and M. Arnott. 1996. Taste and odor of potential potable water contaminants; *Water Research,* 30(2):331-340.

Zhou, E., and R.L. Crawford. 1995. Effects of Oxygen, Nitrogen, and Temperature on Gasoline Biodegradation in Soil. *Biodegradation*, 6(2):127-140.

Zytner, R.G. 1994. Sorption of Benzene, Toluene, Ethylbenzene and Xylenes to Various Media. *Jour. Hazardous Materials*, 38(1):113-126.

ABBREVIATIONS

1, 1, 2-TCA	1, 1, 2-trichloroethane
a	outer radii
α_i	activity of the compound i in liquid
γ	correction factor
β_A	area factor
β_L	length factor
A	contact area
ABS	acrylonitrile-butadiene-styrene
AC	asbestos-cement
AWWA	American Water Works Association
AwwaRF	Awwa Research Foundation
b	inner radii
BTEX	benzene, toluene, ethylbenzene, and xylenes
C_o	concentration of contaminant surrounding the pipe
C_{bulk}	bulk concentration in soil pore water
$C_{m,w}$	maximal solubility in water
C_{pw}	concentration of contaminant in pipe water
C_v	concentration in the vapor phase
$C_{v,m}$	maximal solubility in the vapor phase
C_w	concentration in water
CI	cast iron (pipe)
CR	neoprene
D_o	diffusion coefficient at 25° C
D	diffusion coefficient at temperatures other than 25° C
DCE	dichloroethylene (any isomer)
DI	ductile iron (pipe)
DIPRA	Ductile Iron Pipe Research Association
EPDM	ethylene-propylene-diene-monomer
f_i	fugacity of the compound in a liquid mixture
$f_{i,pure\ liquid}$	fugacity of the compound in the pure organic liquid
FKM	fluoroelastomer rubber
GC	gas chromatography
gpm	gallons per minute
HDPE	high density polyethylene pipe
I.D.	inside diameter of pipe

L_0	original (starting) pipe thickness
ℓ	pipe wall thickness
L_C	length of contaminated pipeline
L_T	total length of pipeline
$L_{s,t}$	thickness of the swollen layer at time t
$L_{g,t}$	thickness of the remaining glass core (the distance of the sharp boundary to the inner wall)
LDPE	low density polyethylene pipe
LUST	leaking underground storage tank
M	total mass of permeated contaminant
MCL	the US EPA maximum contaminant level, an enforceable standard
MCLG	the US EPA maximum contaminant level goal
MDL	analytical method detection limit
MJ	mechanical joint
MTBE	methyl tert-butyl ether
NAPL	non-aqueous- phase liquid
NBR	nitrile rubber
NIR	near infrared
NIST	National Institute of Standards and Technology
O.D.	outside diameter of the pipe
P_m	steady state permeation rate of the contaminant
PC	permeation coefficient
P_i^0	saturated vapor pressure of i in the pure organic liquid
P_m	steady state permeation rate
P_{mp}	predicted steady permeation rate for the wall thickness of l_p
P_{ms}	known steady permeation rate for the wall thickness of l_s
PB	polybutylene
PC	principal component
PCA	principal component analysis or perchloroethene (1,1,2,2-tetrachloroethene)
PVC	polyvinyl chloride (pipe)
PCE	perchloroethylene (tetrachloroethene)
PE	polyethylene
PEG	polyethylene glycol
PEX	cross-linked polyethylene
PLS	partial least squares
PTFE	polytetrafluoroethylene
$Q(t)$	mass of contaminant permeating through unit surface area of a polymer
q	flow rate
RPD	relative predictive determinant

S	solubility
SC	service connection
SBR	styrene-butadiene-rubber
SEL	standard error of laboratory data
SEP	standard error of prediction
T	breakthrough time
t	time
T_L	time lag
TCA	trichloroethene (any isomer)
TCE	trichloroethylene
V	volume of water in the pipe
v	average water flow velocity
V_i	vapor pressure of i equilibrated with a liquid mixture
x_i	mole fraction of i in the mixture
VOC	volatile organic compound

IWA Publishing's authorised EU representative for General Product Safety Regulations is Diane D'Arras, 15 rue Duret, 75116 Paris, France, e-mail: safety@iwap.co.uk.

Printed and bound by CPI Group (UK) Ltd, Croydon, CR0 4YY
17/09/2026
02218813-0005